Science within Art

Science within Art

Lynette I. Rhodes

PUBLISHED BY THE CLEVELAND MUSEUM OF ART

Lenders to the Exhibition

George Levitine, Silver Spring, Maryland
Cleveland Health Sciences Library
The Garden Center of Greater Cleveland

Cover: Rustic Platter [8]
Title page calligraphy by William E. Ward.

Copyright 1980 by The Cleveland Museum of Art
11150 East Boulevard, Cleveland, Ohio 44106
Library of Congress Catalog Card No: 79-93193
ISBN: 0-910386-57-9

Distributed by Indiana University Press, Bloomington, Indiana 47401

Contents

6 Foreword

7 Introduction

Part One

9 Life Sciences

    Anatomy   *Objects*   1- 6

    Nature Studies   7-13

Part Two

30 Material Sciences

    Pigments   14-19

    Acids   20-24

    Glass   25-29

    Lacquer   30

    Ceramics   31-38

    Metals   39-46

69 Conclusion

70 Selected Bibliography

Foreword

The subject of this exhibition is broad — the interactions between science, technology, and art. Its implications and ramifications are enormous, and neither the exhibition nor the catalog claim to be more than an introduction to a complex field.

In a sense this effort is interdisciplinary in nature, bringing together as it does important strands from both art and science. It was inspired by an exhibition held at The Massachusetts Institute of Technology in 1978 entitled Aspects of Art and Science — a highly provocative show, conceived and implemented by Cyril Stanley Smith, a distinguished materials scientist and professor emeritus at the Institute. Art objects for the Cleveland exhibition, most of which come from the Museum's permanent collection, were selected to demonstrate the manner in which processes and techniques developed by the artist/craftsman often predated and stimulated subsequent technological and scientific advances.

A year ago Lynette Rhodes, who was then a staff member of the Department of Art History and Education, assumed the responsibility of organizing this new exhibition, which is based on many of the ideas of Cyril Stanley Smith and on the show which he assembled at a technical university — but which is intended now for an art museum environment and addressed to its audience.

The exhibition Ms. Rhodes prepared offers a new perspective on objects from the Museum's permanent collection. The organization of this show, which is reflected in the catalog, is not based on historical or stylistic categories but according to material and technical considerations. Such considerations, transferred in detail to the printed page, sometimes inhibit or discourage the general reader — and so we have attempted to present the expository sections of this catalog as lucidly and concisely as possible.

No exhibition can be the work of only one person and this is the case with Science within Art. Marianne Doezema worked diligently on various aspects of the project and saw to it that the catalog was placed in production; Katherine Solender gave Ms. Doezema valuable assistance. Additional assistance was provided by June Salm, Joy Walworth, Bernice Spink, Sally Goodfellow, Elinor Pearlstein, Amy Levine, and Marlene Goldheimer. Museum curators who assisted in answering questions on specific pieces include Henry Hawley, Arielle Kozloff, Philippe Verdier, and Dorothy G. Shepherd. Technical advice was given by Frederick L. Hollendonner, Ross Merrill, Delbert L. Spurlock, and Hermine Altmann of the Museum staff. Richard T. Isaacson, librarian at The Garden Center of Greater Cleveland and Glen Jenkins, rare book librarian/archivist in the Historical Division of the Cleveland Health Sciences Library assisted in several matters relating to the exhibition. Technical guidance in the preparation of the catalog was given by Nicholas C. Hlobeczy and David Heald of the Museum's photography studio; William E. Ward and Joseph L. Finizia carefully installed the exhibition. James A. Birch ably coordinated the exhibition during its final stages.

But the major spur to the Museum's organization of the exhibition was the initiative of Director Sherman E. Lee. He recognized a chance for an unusual show and acted decisively in helping us realize the potential of this exhibition.

Gabriel P. Weisberg
Curator of Art History and Education

Introduction

Science and art, in our own day, tend to be thought of as fundamentally separate: one, an analytical tool; the other, the result of an elusive creative process. Actually, however, they are sister reflections of the human imagination — of the ability to explore and interpret, to respond to and shape the realities both within and outside of ourselves.

In earlier civilizations scientist and artist were often one person. The artisan who fashioned an object from fire-hardened clay was also a scientist learning to understand the properties of his materials, and a technologist using these properties to achieve a definite end. Art itself often led the way to insights and discoveries that science later acted upon. Such practice in ancient times as the polishing of mirrors or the cutting of accurate facets on gems to produce a more decorative glitter, contained the germ of later optical devices. The use in the Middle Ages of metal oxides to make stained glass windows and colored enamels ultimately resulted in the chemist's borax bead test. The list of technical discoveries that have evolved from artistic activity is enormous. Excavations at many Middle Eastern archaeological sites show that artistic purposes have indeed often preceded practical ones. Fire-hardened clay figurines, for example, predate the fired pots found at these sites. The copper dress ornaments and beads made in the seventh millennium BC and found in the Chatal Huyuk in Anatolis and at Ali Kosh in Iran considerably preceded the use of copper for weapons. The first uses of ceramics and metals were apparently for magical, religious, and aesthetic rather than for functional purposes.[1]

During the Renaissance, and in fact until the seventeenth century, science and art were closely allied — looked upon as twin aspects of learning. In their study of optics and experimentation with perspective, light, and shadow, for instance, artists such as Alberti, Brunelleschi, and Piero della Francesca exemplify the Renaissance search for scientific truths that would give an underlying order to our visual experience. The idea that art and science are an interlocking endeavor fusing sensual experience with reasoned analysis is intensely evident in the works of Leonardo da Vinci (1452-1519). In a treatise on painting he spoke of the peril of divorcing science from art: "Those who are in love with practice without science are like a sailor who gets into a ship without a rudder or compass and who can never be certain where he is going." And he warned that theoretical knowledge by itself is insufficient: "But to me it seems that all sciences are vain and full of errors that are not born of experience . . . that is to say, that do not at their origin, middle or end pass through any of the five senses."[2]

By the end of the seventeenth century, as a result of the writings of such scientists as Galileo, Kepler, and Newton, who had rejected the medieval doctrines that had grown from the forcible union of certain aspects of Greek natural philosophy with Christian dogma, new foundations for the physical sciences were laid based on mathematics and rigorous empirical testing.[3] During the next three hundred years science and art became increasingly autonomous and have moved along increasingly divergent paths. But strong currents of influence and collaboration between the two have continued to flow.

In our own century, for example, Einsteinian physics has had a profound impact on the arts. Wassily Kandinsky (1866-1944), describing in his autobiography some of the factors that underlay his developing a new style of nonobjective painting, wrote: "A scientific event cleared my way of one of the greatest impediments. This was the further division of the atom. The crumbling of the atom was to my soul like the crumbling of the whole world. Suddenly the heaviest walls toppled. Everything became uncertain, tottering and weak. I would not have been surprised if a stone had dissolved in the air in front of me and become invisible. Science seemed to me destroyed; its most important basis was only a delusion, an error of the learned, who did not build their godly structures stone by stone with a steady hand in transfigured light, but groped at random in the darkness for truth and blindly mistook one object for another."[4]

And Naum Gabo (b. 1890), one of the earlist and best-known constructivist sculptors, who studied mathematics, physics, chemistry, and engineering at the University of Munich from 1909 to 1912, explained how current scientific theory was to influence sculpture: "Older sculpture was created in terms of solids; the new departure was to create in terms of space."[5]

This exhibition focuses primarily on that area where science and art most intimately mesh: the decorative arts. The objects here were assembled for a double purpose: to give aesthetic pleasure and to illuminate the role which art has played in the discovery of material properties. Long before man could articulate the information in scientific terms, he had a vast working knowledge of the behavior of materials under chemical, thermal, and mechanical treatment. Proof of this can be found in artifacts from the seventh millennium BC — well-shaped, decorated pottery from Chatal Huyuk; from 2600 BC — gold jewelry from the royal graves at Ur; and in ceramic, stone, and metal art from ancient Egypt and Persia. In fact almost all the materials used in engineering prior to the twentieth century were known in the decoractive arts before 2000 BC.[6] Today, as Cyril Stanley Smith points out, chemistry, solid-state physics, and in particular the new and growing field of material sciences are making it clear that "the qualities of the materials to which the artist responded sensually are inherently the same as the properties and structures that are measured and explained by the scientists."[7]

Ideally, the viewer's visual pleasure in these objects will be enriched by a heightened awareness of the materials from which they were wrought — how they were understood and used, and how they are integral to the artistic whole. "Works of art," says Professor Smith, "like everything else, are composed of matter, and the matter is always subservient to form. Nevertheless, the hints of larger meaning carried by their shapes and textures are merely the upper levels of a hierarchy of interdependence that stretches downward through the nature of materials."[8]

1. Cyril Stanley Smith, "Art, Technology and Science: Notes on Their Historical Interaction," in *Perspectives in the History of Science and Technology*, ed. Duane H. D. Roller (Norman: University of Oklahoma Press, 1971), pp. 134, 137.
2. Leonardo da Vinci, *The Literary Works of Leonardo da Vinci: Compiled and Edited from the Original Manuscripts by Jean Paul Richter*, vol. 1, 2nd ed. rev. and enl. by Jean Paul Richter and Irma A. Richter (London: Oxford University Press, 1939), p. 119.
3. Jacob Bronowski, "The Discovery of Form," in *Structure in Art and in Science*, ed. Gyorgy Kepes (New York: George Braziller, 1965), p. 57; Hedley Howell Rhys, ed., *Seventeenth Century Science and the Arts* (Princeton: Princeton University Press, 1961), pp. 4-6. For a general history of European science see S. F. Mason, *Main Currents of Scientific Thought: A History of the Sciences* (New York: Henry Schuman, 1953).
4. Wassily Kandinsky, "Rückblicke" [Reminiscences] (Berlin: Herwarth Walden, 1913); English trans. Robert L. Herbert, ed., *Modern Artists on Art* (Englewood Cliffs, N. J.: Prentice-Hall, 1964), p. 27.
5. Peter Blanc, *The Artist and the Atom*, from the Smithsonian Report for 1951, no. 4082 (Washington, D. C.: United States Government Printing Office, 1952), p. 435.
6. Cyril Stanley Smith, "Matter Versus Materials: A Historical View," *Science* 162 (November 1968): 637-38; idem, *Aspects of Art and Science*, exh. cat. (Cambridge: Margaret Hutchinson Compton Gallery, Massachusetts Institute of Technology, 1978), p. 5.
7. Smith, *Aspects of Art and Science*, p. 7.
8. Ibid.

1 Life Sciences

The botanical artist, artist-naturalist, or artist-anatomist is faced with a constant dilemma: is he the servant of science or of art? In actuality, he must learn to serve both masters. The greatest of these painters and draftsmen, who sought to realistically record the world of plants and animals, have been those who found beauty in truth. They understood their subjects scientifically, but saw and described them with the eye and hand of the artist.

Man's initial interest in the uses and habits of plants and animals was probably stimulated by their value as food and his own concern for self-preservation, but the presence of flowers in Neanderthal graves strongly suggests that man's investigation of the world around him was also stimulated by his capacity for aesthetic enjoyment.[1] This dual interest in nature, based on both necessity and pleasure, continued to motivate man to increase his understanding of his environment — an understanding that has resulted in a vast body of drawings and paintings with combined scientific and aesthetic appeal.

The life sciences deal with all of the physiochemical aspects of life — that is, they investigate the physical and chemical phenomena common to all living things, and also the actions and functions that distinguish living from nonliving things. Explanations and illustrations of man's interest in the characteristics and functions of plants, animals, and indeed himself, may be found among his earliest written records and on ancient artifacts. As with many of the sciences, however, the beginning of formal biological investigation is associated with the ancient Greek scholars who concerned themselves with science in its broadest sense. Aristotle (384-322 BC), a student of Plato, attempted the first system of animal classification and also inquired into the nature of plants. Theophrastus (370-ca. 285 BC), a student of Aristotle and author of the oldest extant botanical treatise, described the forms, structures, processes, functions, and therapeutic uses of over 500 plants, and began to establish a scientific terminology for botany. Later, Galen, a Greek physician who practiced in Rome during the mid-second century AD, wrote extensively on human anatomy. His theories dominated medicine for centuries, despite the fact that his observations were based on animal rather than human research and contained many errors.[2]

Following this initial age of inquiry, biological research diminished as Greece and Rome ceased to be the centers of learning. During the medieval period, however, scientific knowledge was kept alive by Arab physicians and philosophers who had acquired the Greek treatises.[3] The Arabic translations were eventually rendered into Latin, and then, sometimes, even into Greek again, and in these guises found their way into the mainstream of western Europe.

During this period, between the classical age and the end of the fourteenth century, general curiosity about the medicinal properties of plants and the nature of animals engendered the writing of many scientific and pseudoscientific manuscripts. Herbals were books containing the names and descriptions of herbs or plants with their properties and virtues.[4] Bestiaries were the zoological counterparts of herbals. Unfortunately, most of these works lacked a sound scientific basis. Herbals were based upon Greek and Latin writings and Arabic commentaries rather than upon direct observations of nature, and bestiaries repeated ill-founded animal myths combined with convoluted Christian theology. The highly stylized plant and animal illustrations found in both added little more than decorative embellishments to the texts.

With the Renaissance came a revival of naturalism, and naturalistically treated flowers, fruits, and animals graced the work of the foremost painters, sculptors, and craftsmen. The understanding of plant structure evident in Leonardo da Vinci's vigorous flower drawings places them among the first truly modern botanical renditions. The delicate watercolor paintings of Albrecht Dürer (1471-1528) also have the exactness of detail found in the best botanical or zoological studies.[5] Both artists are famous for their exquisite studies of domestic as well as exotic animals. The nature scenes and animals in the pottery designs of Bernard Palissy (1510-1589) were actually cast from real specimens, and their anatomical accuracy and humorous juxtapositions often amuse naturalists. The study of anatomy was revitalized toward the end of the fifteenth and the beginning of the sixteenth century by the introduction of human dissection into the curricula of medical schools and art academies. The first anatomic illustrations had appeared in Johannes de Ketham's *Fasciculus Medicinae* in 1491 and were intended for physicians' use; these, however, were simply schematic representations.[6] Progress in scientific anatomy continued to be hindered by the perpetuation of Galenic-Arabic canons and attitudes. Yet anatomical illustration was stimulated by the draftsman's need for an understanding of body mechanics in order to render the human form precisely. During the sixteenth century the practice of studying and drawing anatomy from nature helped to eliminate many traditional misconceptions, to broaden anatomical knowledge, and to improve anatomic illustration. In 1543 the publication of the well-illustrated and detailed *De Corporis Humani Fabrica* by Andreas Vesalius (1513/14-1564) revolutionized the teaching of anatomy, influenced the style of anatomic illustration, and provided the foundation of modern anatomical science.[7]

Leonardo contributed further to the growing understanding of human mechanics. His detailed anatomical studies, based on numerous dissections, were so advanced for their time that some of their details were not appreciated until a century later. In addition, Leonardo introduced the study of comparative anatomy by dissecting animals and comparing their structure with that of man. The noting of structural similarity, or homology, later became important to the study of evolution.[8] By the end of the sixteenth century the increased availability of paper, the perfection of wood-engraving techniques, and the acceptance of steel and copper engraving as media for reproduction had revolutionized the crafts of the nature illustrator.[9] In addition, his choices of imagery dramatically increased as explorers introduced to Europe their recordings of collections of specimens from the rich variety of wildlife and natural resources found in the New World. In 1735 Carolus von Linnaeus revolutionized the world of natural science with the publication of *Systema naturae*, which provided what is known as the binomial system of nomenclature, the system of animal and plant classification that is in use to this day. An immediate consequence was the obsolescence of all previous botanical works and a concentration of study and effort to revise the older works and produce new botanical studies based on the Linnaean system. This necessitated a new type of illustration, emphasizing scientific accuracy and visual realism, which could be produced only by draftsmen who combined artistic skill with scientific botanical knowledge.[10]

By the eighteenth century, interest in useful plants had to a great extent yielded to a passion for the beautiful, and artists were sought to record the rare flowers grown in the gardens of the wealthy. This demand persisted into the nineteenth century; Pierre-Joseph Redouté (1759-1840), perhaps the most celebrated painter in the history of botanical art, illustrated several magnificent folios documenting the formal gardens cultivated by Josephine Bonaparte at Malmaison.[11]

There also emerged a new breed of scientific artist, the artist-naturalist. With a talent for acute observation and the ability to capture the essential beauty of the plants and animals they so diligently sought and carefully drew, these pioneers traveled to the wilderness and achieved in their work a fresh balance between art and science. Among the most able of those who cataloged the natural history of the New World were Maria Sybilla Merian (1647-1717), a German woman who visited Surinam to paint the insects and smaller vertebrates of this area in their natural surroundings, and John James Audubon (1785-1851), an American who painted more than 435 species of North American birds in their native habitats and daily activities.

As the study of the life sciences became rapidly more sophisticated, the camera, for the most part, began to replace the artist's brush or chalk as the principal means of scientific documentation. Yet the artist's delight in the structure and characteristics of the natural world and its inhabitants did not diminish.[12] The flower or animal paintings by some modern artists, such as Georgia O'Keeffe (b. 1887), offer much more than a visual record of the variety of nature. They train one to look beyond the obvious, to the structural miracles of cell and tissue which are found in each living thing, and to the subtle beauties of color, pattern, and texture which are foundations of art as well as components of nature.

1. Cyril Stanley Smith, "Art, Technology, and Science: Notes on Their Historical Interaction," in *Perspectives in the History of Science and Technology,* ed. Duane H. D. Roller (Norman: University of Oklahoma Press, 1971), p. 134.

2. Ludwig Choulant, *History and Bibliography of Anatomic Illustration in Its Relation to Anatomic Science and the Graphic Arts,* trans. Frank Mortimer (Chicago: University of Chicago Press, 1920), pp. 26-28. See also George Sarton, *A History of Science: Ancient Science Through the Golden Age of Greece* (Cambridge: Harvard University Press, 1952).

3. Soon after the time of Alexander the Great (356-323 BC), Greek schools began to be founded in Syria. From these centers the teachings of the Greek philosophers were handed on in Persia, Arabia, and other countries. Agnes Arber, *Herbals: Their Origin and Evolution, A Chapter in the History of Botany 1470-1670* (Cambridge: At the University Press, 1938), pp. 2-4.

4. Ibid., pp. 12-14.

5. Wilfrid Blunt assisted by William T. Stearn, *The Art of Botanical Illustration* (London: Collins, 1950), pp. 23-30.

6. Choulant, *Anatomic Illustration*, p. 27.

7. Ibid., pp. 30-33, 169-99; Choulant thoroughly considers anatomic illustration from two viewpoints: the influence of graphic arts on anatomic science and the influence of anatomic science on the graphic arts.

8. *The Drawings of Leonardo da Vinci*, with Introduction and Notes by A. E. Popham (New York: Reynal & Hitchcock, 1945), pp. 58-65. See also James Ackerman, "Science and Art in the Work of Leonardo," in *Leonardo's Legacy: An International Symposium,* ed. C. E. O'Malley (Berkeley and Los Angeles: University of California Press, 1969), pp. 205-25.

9. A. M. Lysaght, *The Book of Birds: Five Centuries of Bird Illustration* (London: Phaidon Press, 1975), p. 17.

10. Gordon Dunthorne, *Flower and Fruit Prints of the 18th and Early 19th Centuries* (Washington, D. C.: Gordon Dunthorne, 1938), p. 3.

11. Blunt, *Botanical Illustration*, p. 176.

12. For an interesting discussion of the parallel interests of scientists and naturalistic landscape painters in nineteenth-century France, see Aaron Sheon, "French Art and Science in the Mid-Nineteenth Century: Some Points of Contact," *Art Quarterly* 34 (Winter 1971): 434-55.

12

1 Pierfrancesco Alberti, Italian, 1584-1638. *A Painter's Academy in Rome*. Engraving, 8-21/32 x 11 inches (22 x 28 cm.). The Metropolitan Museum of Art, The Elisha Whittelsey Collection, The Elisha Whittelsey Fund. Photoreproduction.

During the Renaissance, young, aspiring artists received their training at academies, which were actually the studios of established masters. There they learned the skills necessary for their careers, especially draftsmanship — which was then considered the basis of all architecture, sculpture, and painting. The structure of the human body became a central course of study as artists sought anatomical knowledge as a foundation for their art.

Pierfrancesco Alberti is perhaps best known for the engraving *A Painter's Academy in Rome* [1]. This view of an artist's workshop illustrates the emphasis that was placed on the study of anatomy. Students work among an assortment of anatomical study aids: plaster casts of body parts, death masks, portrait busts, a skeleton, and a corpse. In the central foreground two young men draw from the skeleton as a third offers advice; meanwhile, to their right, two other students model figures in clay. At the far left the teacher reviews a drawing of eyes, probably copied from an artist's primer, and near the rear door another student sketches a plaster cast of a leg. A dissection and cadaver study are being conducted in the back of the room.

1 Pierfrancesco Alberti, *A Painter's Academy in Rome*.

The young draftsmen also study mathematics. A boy near the door uses a ruler to plot an architectural background on a canvas, and a group of students in the center of the room discuss geometry. The latter is a visual parody of Raphael's famous painting *The School of Athens*, in which Euclid and Ptolemy, amidst a similar arrangement of figures, also discuss geometry. The three pictures on the wall represent appropriate subjects for painting: a landscape, a portrait, and a Biblical scene.[1]

1. Most of the material in this entry comes from Edward J. Olszewski with Jane Glaubinger, *The Draftsman's Eye: Late Italian Renaissance Schools and Styles* (Cleveland: The Cleveland Museum of Art, 1980), cat. no. 31.

2 Giovanni Battista Franco, Italian, ca. 1510-1561. *Half-Length Skeleton in Profile*. Pen and brown ink on paper, ca. 1545, 9-1/4 x 6-5/16 inches (23.4 x 16.1 cm.). Gift of Mr. and Mrs. Claude Cassirer. CMA 64.381

3 Giovanni Battista Franco, Italian, ca. 1510-1561. *Full-Length Skeleton from the Back*. Pen and brown ink on paper, 16-9/16 x 6-13/16 inches (42.2 x 17.3 cm.). Gift of Mr. and Mrs. Claude Cassirer. CMA 64.380, 64.380a

It was customary during the Renaissance for anatomy manuals to be illustrated with drawings that were artistically beautiful as well as scientifically correct, and figures were often shown in poses that expressed the perfection of nature or a human ideal.

A multiple view of a skeleton, showing front, side, and back on a single page, was quite common. This composition resulted in a well-organized drawing that was artistically interesting and at the same time most useful for instructional purposes. In Franco's drawings [2, 3] the skeletons seem to hold their poses like marionettes standing against the blank page.[1] The bony hand resting on the shoulder of the full-length figure actually belongs to the drawing of the skeleton in profile, and the hand and forearm at the right of the larger image are part of yet a third figure, indicating that the page once extended further to the right. Because of surface abrasion, the right arm of the full-length figure is barely visible, but under ultraviolet light it may be seen to extend horizontally to the edge of the paper, then bend sharply at the elbow so that the fingers descend from the top of the page to touch the right shoulder. This pose must be one that Franco drew directly from a skeleton, because the right scapular plate, again more visible under ultraviolet light, is seen from the edge and foreshortened, as though it had been badly displaced during the repositioning of the skeleton's arm.[2]

The drawings also contain other anatomical exaggerations and inaccuracies. In the half-length profile, for example, the spinal column is virtually straight and the centra (body) of the individual vertebrae appear too heavily ridged. The sternum and lower ribs are also distorted, and it has been suggested that the pelvis does not even belong to the same set of bones as the rest. Similar errors occur in the drawing of the full-length skeleton.[3] These discrepancies indicate that Franco was not especially interested in scientific anatomy or the exact form and movement of the human body. Instead, he

13

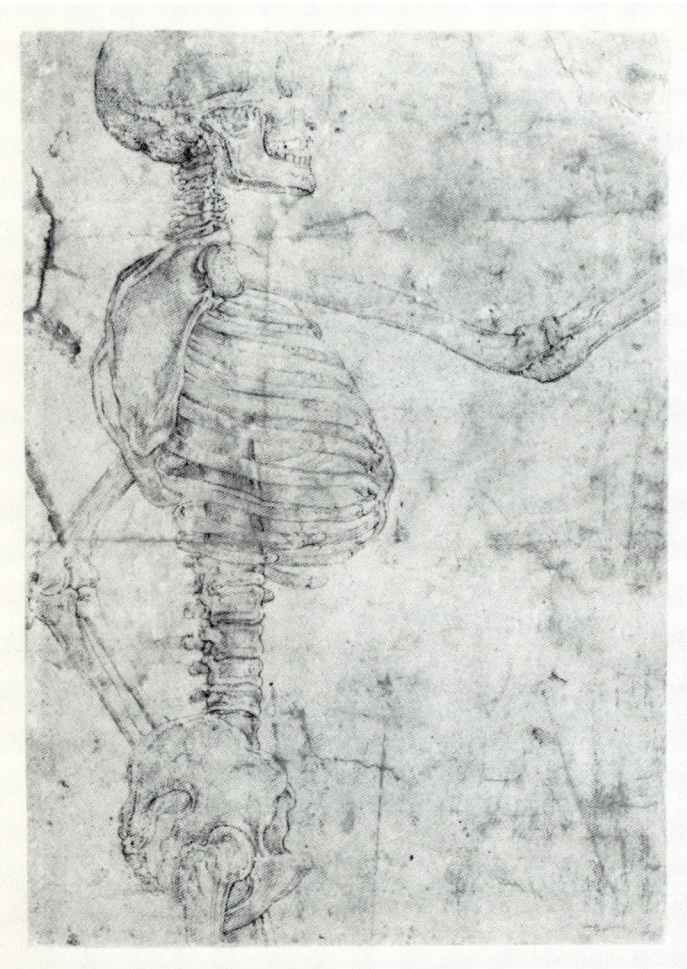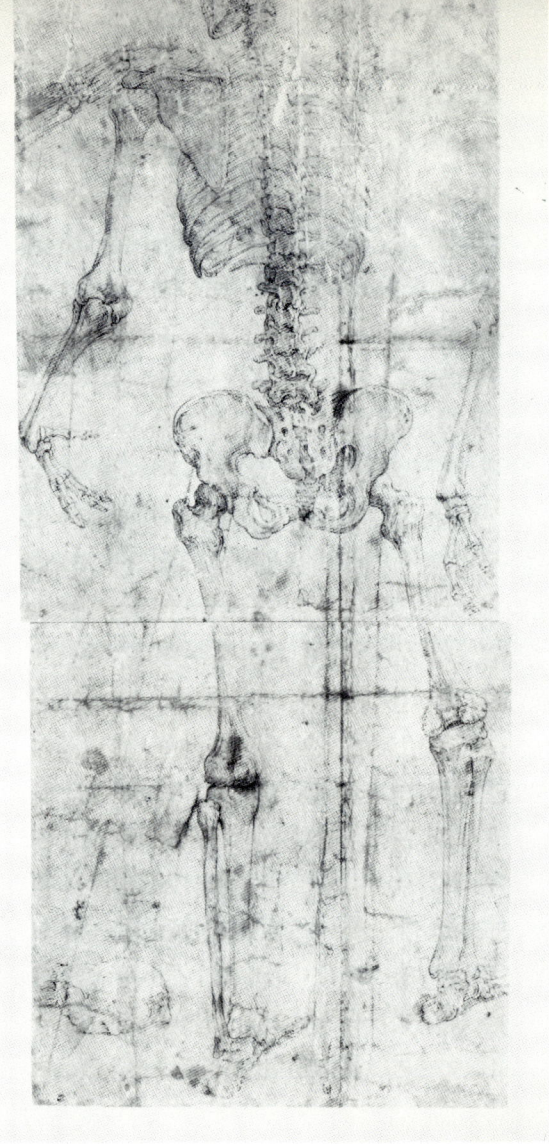

was concerned with artistic anatomy, the idealized drawing of anatomic structures that reveals the delicate intricacy of the human skeleton and the beautiful shapes of the bones themselves.

1. There are six drawings of skeleton and bone subjects by Giovanni Battista Franco in the Cleveland collection. All are pen and brown ink on paper darkened to a tan: *Lower Half of Skeleton from the Front*, 11-1/8 x 4-15/16 inches (28.3 x 12.6 cm.); *Arm Bones*, 4-5/8 x 14-7/16 inches (11.8 x 36.7 cm.); *Rib Cages*, 4-9/16 x 9-3/8 inches (11.6 x 23.8 cm.); *Torsos with Rib Cages*, 4-11/16 x 13-1/16 inches (11.9 x 33.3 cm.); The Cleveland Museum of Art, nos. 64.382, 64.379, 64.383, 64.378. Franco, a painter, engraver, and book illustrator, was born in Venice but worked primarily in Rome and Florence. The half-length skeleton profile seen here was incorporated by Franco into the engraving *Design of Skeleton, Skulls, and Bones*, 18-1/8 x 12-3/4 inches (46 x 32.4 cm.), The Albertina, Vienna. See L. S. Richards, "Drawings by Battista Franco," *The Bulletin of The Cleveland Museum of Art* 52 (October 1965): 107-12. See also Edward J. Olszewski with Jane Glaubinger, *The Draftsman's Eye: Late Italian Renaissance Schools and Styles* (Cleveland: The Cleveland Museum of Art, 1980), cat. nos. 48, 49.
2. L. S. Richards, "Giovanni Battista Franco's Anatomical Drawings in Cleveland," *Journal of the History of Medicine and Allied Sciences* 20 (1965): 407.
3. Thomas R. Forbes, "Addendum: Franco's Osteology," *Journal of the History of Medicine and Allied Sciences* 20 (1965): 409.

2. Giovanni Battista Franco, *Half-Length Skeleton in Profile*.

3 Giovanni Battista Franco, *Full-Length Skeleton from the Back*.

4 Attributed to Bartolommeo Torre da Arezzo (or Aretino), Italian, ca. 1529-ca. 1554. *Studies of a Flayed Man:* one figure, torso and arm (verso, A); two figures, pelvis and legs (recto, B). Pen and brown ink, brown wash over black chalk, incised, 1554, 16 x 10-7/8 inches (40.6 x 27.6 cm.). Purchase, L. E. Holden Fund. CMA 75.26

Bartolommeo Torre da Arezzo based his *Studies of a Flayed Man* [4] on direct observations; yet, like Franco, he was more interested in the artistic idealization of the human body than in scientific accuracy. Nonetheless, his drawings of muscles reflect an age when, although each applied his new knowledge differently, the artist and scientist were united more than ever in the exploration of the human body.

The studies, one of a torso and arm and the other of the pelvis and legs of two figures, were executed on both sides of a single sheet of paper. In the first study (A) the abdominal, serratus, pectoralis, and deltoid muscles of the upper torso are easily identified, as are the biceps, triceps, brachialis, and long supinator of the right arm. The shaded areas on either side of the abdominal cavity suggest that the cadaver was dissected and then reconstructed.

The second study (B) examines the muscle structure of the leg from both a frontal and three-quarter view, clearly illustrating the arrangement of muscles that flank the leg bones. Here, as in the other drawing, the actual shapes of the muscles are generalized, creating abstract forms that are rhythmically interwoven.

By showing figures that are only partially dissected, Torre graphically contrasts the smooth rippling of the skin surface with the complex, layered structure of the muscles beneath. The figures dangle in space because it was customary to suspend cadavers and skeletons by the back of the neck while they were being studied and drawn. The incised lines visible on both drawings, and the cape draped over the left shoulder in the torso drawing, suggest that these images were perhaps used as engraved text illustrations.[1]

1. This discussion of Torre's work is based primarily on Edward J. Olszewski with Jane Glaubinger, *The Draftsman's eye: Late Italian Renaissance Schools and Styles* (Cleveland: The Cleveland Museum of Art, 1980), cat. no. 66.

16

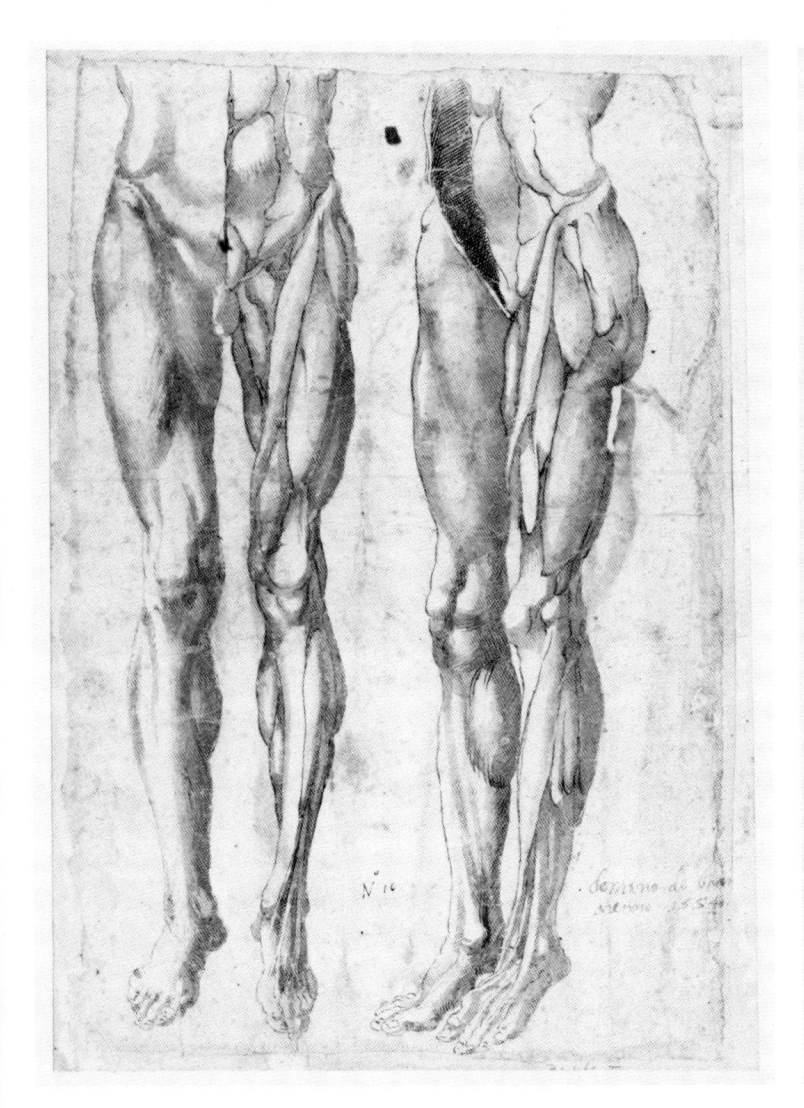

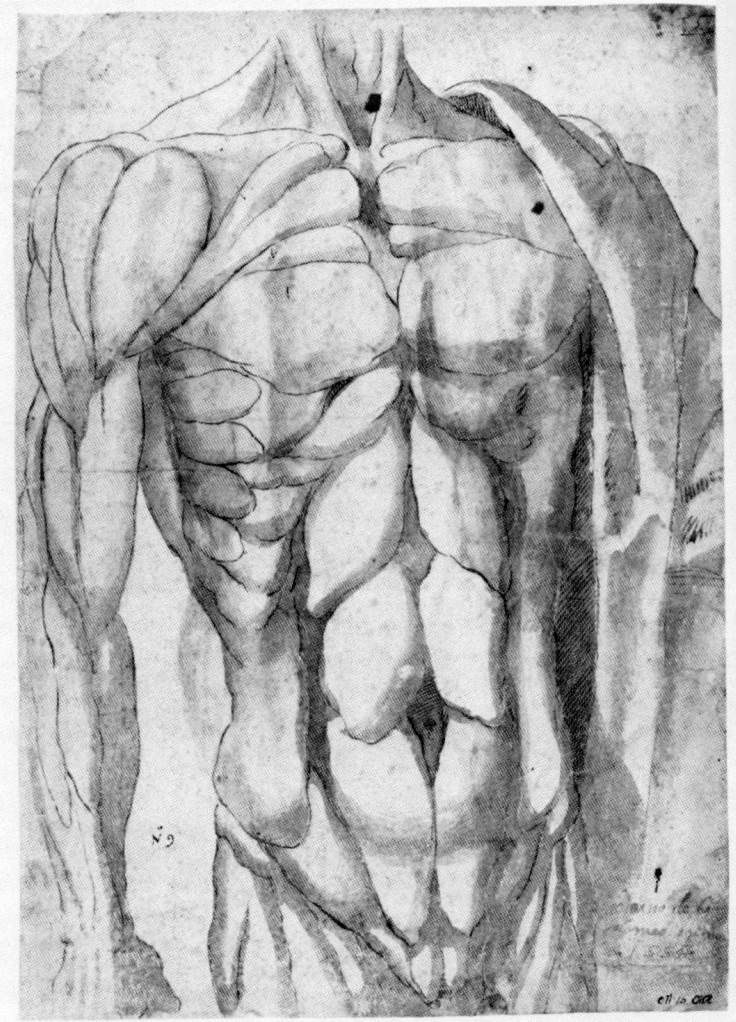

4 A, B Attributed to Bartolommeo Torre da Arezzo, *Studies of a Flayed Man.*

5 Gerard de Lairesse (draftsman), Dutch, 1640-1711; Peter and Philip Van Gunst (engravers), Dutch. *Divers Muscles on the Superior and Fore Part of the Trunk of the Body.* Copperplate engraving, 18-7/8 x 12-1/2 inches (48 x 31.7 cm.), in William Cowper, *The Anatomy of Humane Bodies with Figures Drawn After the Life by Some of the Best Masters in Europe . . .* (Oxford, 1697; 2d ed. rev. by C. B. Albinus, Leyden: Joh. Arn. Langerak, 1737), table 20. Cleveland Health Sciences Library, Historical Division.

6 Nicolas Henri Jacob, French, 1782-1871. *Coupe Verticale de L'encéphale.* Lithograph printed in colors, 16-3/4 x 12-1/4 inches (42.5 x 31.1 cm.), in Jean-Baptiste-Marc Bourgery, *Anatomie Descriptive ou Physiologique,* vol. 3 (Paris: C. A. Delaunay, 1844), pl. 22. Cleveland Health Sciences Library, Historical Division.

The tradition of anatomic illustration that began with Vesalius in the sixteenth century combined scientific exactness with an artistic attitude. It stressed the development of an anatomical norm based on accuracy and raised it to an artistic mode of representation. This approach, which has continued to dominate anatomical representation to this day, was influenced at times by prevailing

5 Gerard de Lairesse (draftsman); Peter and Philip Van Gunst (engravers), *Divers Muscles on the Superior and Fore Part of the Trunk of the Body.*

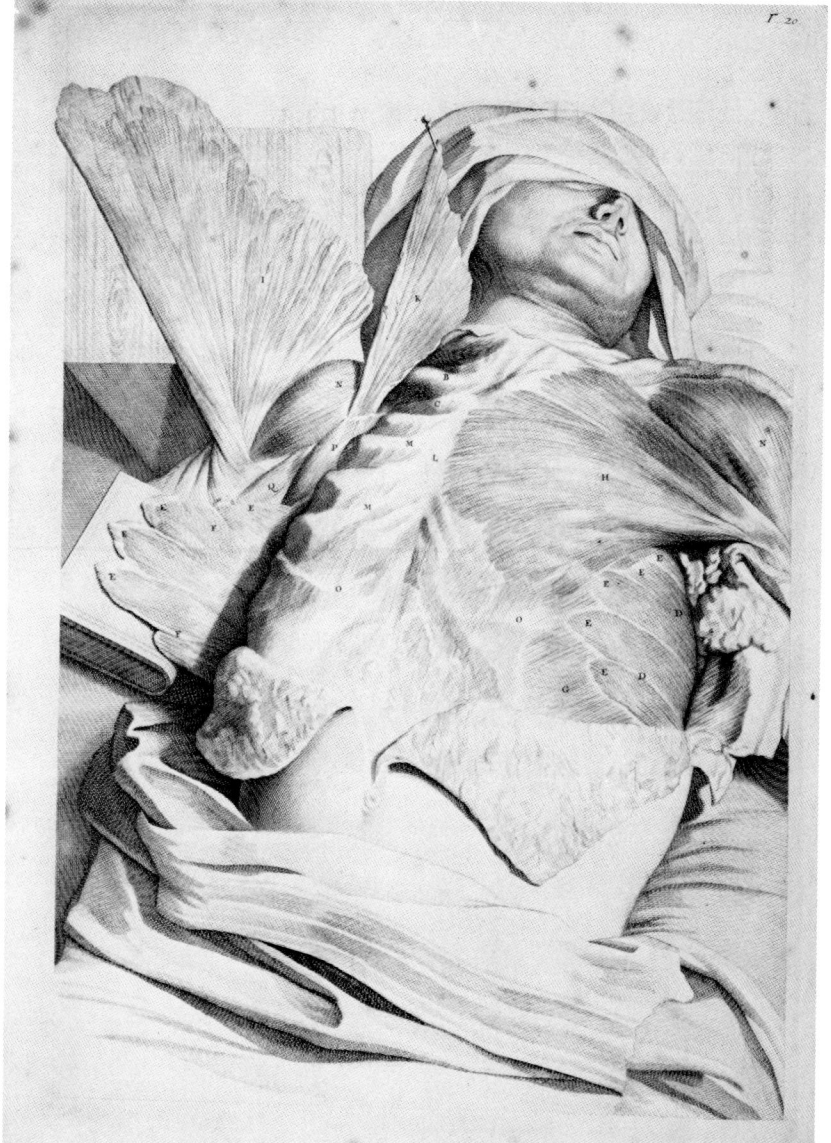

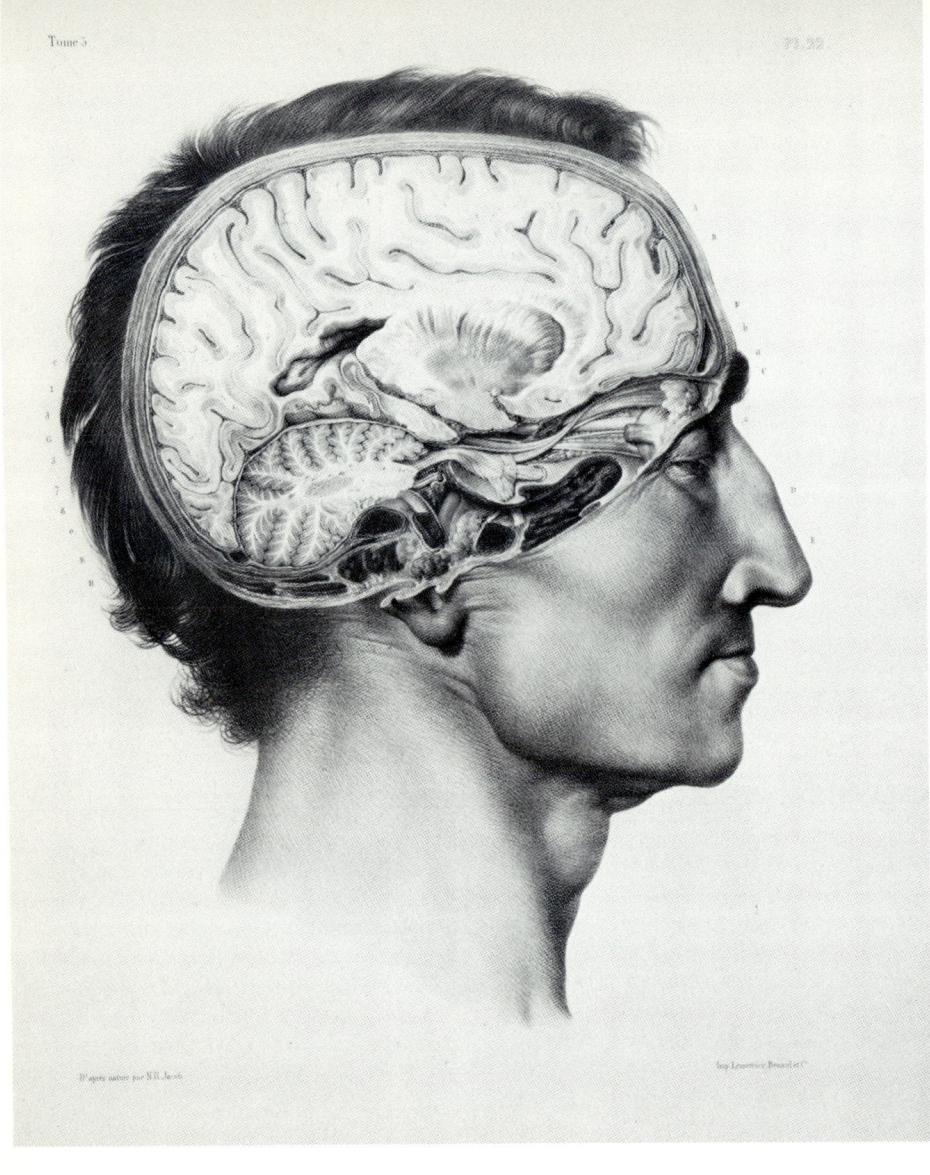

fashions or sentiments, but continually reflected a growing body of knowledge about human structure and function — knowledge which expanded with the development of such new research techniques as the microscopic study of tissue structure, the study of anatomy by means of frozen specimens, and cross-section anatomy. Scientific advances, of course, complemented technological advances that had become available to the artist, including the introduction of copperplate and steel engraving, lithography, photography, and other reproductive processes.[1]

Two anatomical atlases, one from the late seventeenth century and the other from the mid-nineteenth century, may be taken as illustrative and representative of the rich tradition of fusing scientific accuracy with artistic skill. *The Anatomy of Humane Bodies with Figures Drawn After the Life by Some of the Best Masters in Europe in 114 Copper-Plates* was first published at Oxford in 1697 by William Cowper, English surgeon and anatomist. Its elegant engravings illustrate specific body parts as well as the total figure. Several are a curious blend of anatomical documentation and allegory. In one of the two skeleton plates the bony figure, holding an hourglass in its left hand, stands in an open portico beside a sarcophagus with the cover removed, and in front of a second with the lid closed. In another illustration a very realistically rendered fly (a common seventeenth-century symbol of man's mortality) alights on an opened, partially dissected abdomen.

6 Nicolas Henri Jacob,
Coupe Verticale de L'encéphale.

The twentieth table, a study of *Divers Muscles on the Superior and Fore Part of the Trunk of the Body* [5], is of a dissected torso drawn from a dramatic, foreshortened angle and lighted with strong highlights and shadows. The flowing drape across the figure's face and its unblemished skin, suggesting a more affected than natural beauty, make this a rather romantic study, despite its anatomical detail.

Most of the studies in this magnificent atlas are correct, within the limits of the artist's capacity to observe, but unfortunately he lacked expert guidance from the anatomist for whom he worked. Muscles are often misplaced and tissue incorrectly characterized.[2] The drawings were of little use to professional anatomists or artists because of these shortcomings; yet they remain an outstanding example of the collaboration between the artist and scientist that marked this period.[3]

The mid-nineteenth century *Anatomie Descriptive ou Physiologique* by Jean-Baptiste-Marc Bourgery contains not only exquisite studies of the human body but also detailed illustrations of surgical techniques and microscopic analyses. Some of the lithographs show the development of cross-section anatomy, which added a whole new dimension to the study of anatomy by artists and was also a valuable aid to practical medicine.

Coupe Verticale de L'encéphale [6], the twenty-second plate in volume three, is a cross-section study of the brain. In addition to examining the right lobe of the brain, the drawing reveals the adjacent structures, including the layers of the scalp and skull, the optic nerve, and the structure of the eye as well as the inner ear. The skin and skeletal structure of the profiled head are defined in subtly shaded flesh tones and deep gray shadows. Bourgery notes in the accompanying text that while it was drawn with great exactness, the study does not represent the head or brain of a specific individual.

1. Ludwig Choulant, *History and Bibliography of Anatomic Illustration in Its Relation to Anatomic Science and the Graphic Arts*, trans. Frank Mortimer, (Chicago: University of Chicago Press, 1920), pp. 41, 33.
2. Ibid., pp. 34, 250-52.
3. Ibid., pp. 252-53. Cowper's *The Anatomy of Humane Bodies* has a curious publication history. Godfried Bidloo, a Dutch anatomist, originally wrote the text, commissioned 105 of the engravings, and published the work in 1685. It was not well received, possibly due to many inaccuracies, and the editors eventually gave 300 impressions of the plates to Cowper. He added nine plates, wrote a new text in English, and published it under his own name with the above title. The original title-page design remains, except that the shield which contained Bidloo's original title and his name now holds Cowper's name and the revised title.

7 Italy, Venice. *Acorus*. Woodblock print, 3-3/4 x 3-3/8 inches (9.6 x 8.7 cm.), in *De Vitutibus Herbarum* (Venice: Giovanni and Bernardino Rosso, 1509). Cleveland Health Sciences Library, Historical Division.

From very early times, a variety of herbs had been used as healing agents, and it is from these purely utilitarian beginnings that scientific, systematic botany evolved. Most early herbalists were physicians who studied botany because of its connection with the medicinal arts, and at first herbs were categorized according to the qualities that made them valuable to man. This began to change only as man's understanding of plants developed into the science of botany and the need for a type of classification based upon affinities within the plant kingdom became apparent.

In addition to medicine, the science of botany is especially indebted to the arts of printing and engraving. Through these processes, the traditional lore recorded in the manuscript herbals was transferred, with little loss of continuity, to the printed herbals that began to appear in the late fifteenth century. The draftsman and the engraver, however, did more than simply record and transmit existing knowledge. With their acute observations and their painstaking attention to detail, they often revealed anatomic structures and qualities that might otherwise have been overlooked by the botanist.[1]

An early, significant example of the printed herbal is *De Vitutibus Herbarum*, first published in Mainz, Germany, in 1484 by Peter Schöffer. Like most early herbals, it was a compilation based upon medieval manuscripts and Greek and Arabic treatises. Schöffer's herbal became quite popular, and as often happened, was soon pirated. Secondary editions and translations appeared in Bavaria, the Low Countries, and Italy.[2] The Italian version of *De Vitutibus Herbarum* in the exhibition is opened to the illustration and discussion of the acorus plant [7], a pungent aromatic herb that is a member of the arum family. As shown in the print, it has thick, creeping rootstocks that send up two-edged, swordlike leaves and a similarly shaped spathe, the large, leaflike structure in

Below
7 *Acorus.*

Below right
8 Manner of Bernard Palissy, *Rustic Platter.* See also cover illustration.

the center of the plant that holds a flower cluster (spadix). A miniature feeding ibis stands beside the representation of the herb in an almost vacant landscape.

As is characteristic of these early books, the line drawing of the plant is quite bold, but lacks the technical refinements and anatomic precision that make botanical illustration a valuable scientific tool. Later, towards the end of the seventeenth century, botany became more scientific. The herbal, with its traditional mixture of medical and botanical lore, was replaced by the exclusively medical *pharmacopoeia* on the one hand, and the exclusively botanical *flora* on the other.

1. Agnes Arber, *Herbals: Their Origin and Evolution, a Chapter in the History of Botany 1470-1670* (Cambridge: At the University Press, 1938), pp. 264-65.
2. Ibid., pp. 20, 268.

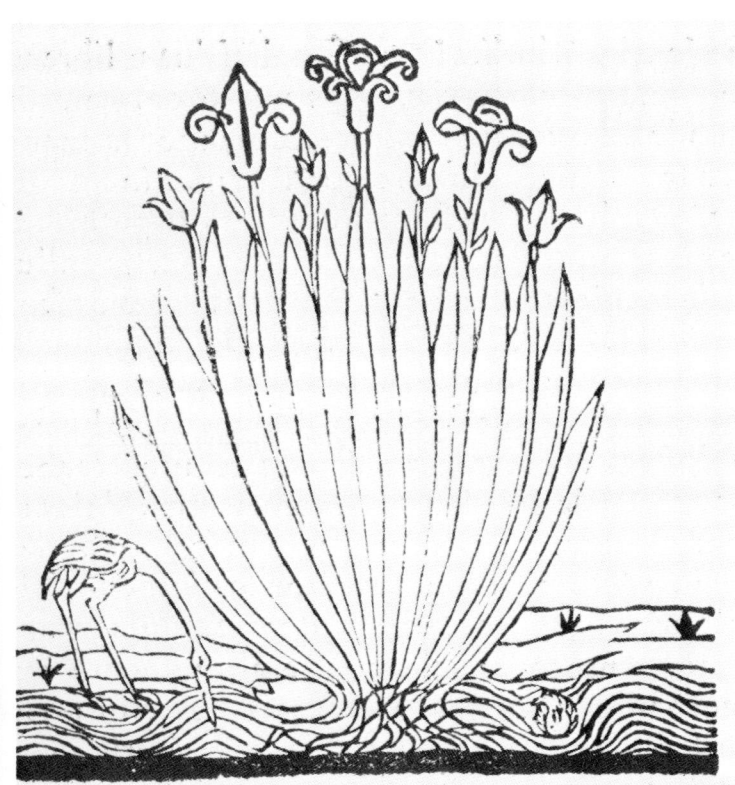

8 Manner of Bernard Palissy, French, ca. 1510-1589. *Rustic Platter.* Lead-glazed earthenware, late 16th century, 20-5/8 x 16 inches (52.4 x 40.6 cm.). Gift of The John Putnam Foundation. CMA 69.106

The Renaissance saw advances in all areas of the natural sciences, and the artist-naturalist, no longer satisfied with the facile rendering of plants and animals, developed an all-consuming thirst for knowledge of the great truths of nature. For Bernard Palissy, science in general was the avenue to this knowledge, and thus he devoted himself to observation and experimentation. He was a theorist of natural science, a propounder of new doctrines in all of its branches, and a pioneer in geological and chemical research. Professionally, however, Palissy was a potter. His geological and chemical studies evolved from his search for the minerals which would produce a perfect "white enamel" pottery.[1] These inquiries, along with his studies of nature as a whole, inspired the distinctive earthenware that is so closely associated with him today.

Palissy began decorating his ware with fanciful arrangements of fish, reptiles, plants, and shells at the time of the Paris basin geological excavations around 1570. During this period Palissy took casts of all the fossil forms he saw in order to study and compare them; and interestingly, the shells in his pottery motifs were based on fossils found at this site. Each of the creatures in his plant and animal designs were also molded directly from the natural object. The choice of subjects varied from piece to piece.[2]

The *Rustic Platter* [8] in the exhibition, which is in the manner of Palissy's work, shows a variety of plants and animals which seem to emerge from and float upon a rippled, "watery" surface. The tableau includes four fish, four snakes, two frogs, two beetles, several salamanders, a crayfish, a butterfly, and a dragonfly. They are carefully placed in naturalistic positions amidst a variety of shells and assorted foliage, including ferns, ivy, bay leaves, and two small, yellow flowers. It is a vital, well-integrated composition, teeming with life; yet each neatly executed specimen also approximates the true appearance of its species. The colors — vivid greens, blues, and yellows, and more somber grays and browns — were applied as glazes, based on lead silicate and colored with metallic oxides.

This trompe-l'oeil serving platter swarming with "live" animals is appealing in both its concept and craftsmanship. It is more important from a scientific point of view, however, because it exemplifies the revolutionary concern with observing, understanding, and recording the natural world that characterized the Renaissance. It also had particular appeal for the decorators of the nineteenth century, when Palissy's work was deeply appreciated and copied.[3]

1. M. L. Solon, *A History and Description of the Old French Faience: With an Account of the Revival of Faience Painting in France* (New York: Cassell and Co., 1903), pp. 30-31. See also Henry Morley, *Palissy the Potter: The Life of Bernard Palissy, of Saintes, His Labours and Discoveries in Art and Science: With an Outline of His Philosophical Doctrines, and a Translation of Illustrative Selections from His Works*, 2 vols. (London: Chapman and Hall, 1852), 1:114, 121, 122-30, 157-71.
2. In 1580 Palissy gave public lectures on geology based on the Paris discoveries, illustrating them with diagrams and experiments. Morley, *Palissy*, 2:178-83; see also Solon, *A History and Description*, p. 33.
3. Palissy had several successors and imitators, but with careful examination of clay body and glaze it is possible to recognize the sharpness of detail and the harmonious color combinations of the work of Palissy himself. Furthermore, a forgery is often betrayed by marbling on the back of the dishes — dry and patchy work, or tinted glazes imperfectly blended. Solon, *A History and Description*, pp. 35-36.

9 Maria Sibylla Merian (painter), German, 1647-1717; Pieter Sluyter (engraver), German. *Poivre d'Inde*. Engraving colored by hand, 13-3/8 x 9-1/8 inches (34 x 23.2 cm.), from *Histoire Générale des Insectes de Surinam et de l'Europe . . .*, 3d ed., vol. 1 (Paris: Buchoz, 1771), no. 55. Dunthorne 205. Gift of The Print Club of Cleveland, In Honor of Arnold M. Davis. CMA 55.320

Maria Sibylla Merian was one of the finest natural-history painters of the late seventeenth and early eighteenth centuries. Her subjects included plants, birds, and reptiles, but she was primarily an entomologist. Her studies of insects and the plants upon which they feed stirred the European imagination, as they introduced the strange inhabitants of distant lands and offered new perspectives on the curiosities found in local gardens.

Because Merian was one of a family of fine artists, her choice of an artistic career was not especially remarkable.[1] But her predilection for entomology, or the study of insects, was unusual for a girl in her day. Apparently her mother had first opposed this study, but then, recalling that during her

pregnancy she herself had enthusiastically collected "caterpillars, shells and stones," she allowed the young Maria to pursue her interest.[2]

In 1679 Merian published the first of three volumes on European insects.[3] This was illustrated with her own engravings, colored by herself and her younger daughter Dorothea. Merian published another illustrated volume, *Metamorphosis Insectorum Surinamensium*, in 1705, after two years of collecting and painting the insects and plants of the Surinam territory along the north central coast of South America. Sixty plates depicted insects in various stages of development amidst their natural surroundings.[4]

Many years after her death, seventy-two of Merian's studies were reproduced as engraved plates in *Histoire Générale des Insectes de Surinam et de l'Europe...*, compiled by the medical botanist Buchoz.[5] *Poivre d'Inde* [9], a plate from this detailed and decorative publication, records two metamorphic processes. A red pepper develops sequentially from an immature, yellow bud at the top of the plant to a firm, ripe fruit in the lower right corner. The cluster of seeds inside one pepper is revealed in a cross-section view. Meanwhile, the transformation of a hawk-moth from a green and yellow caterpillar feeding on the plant's lower leaves to a red, brown, and blue adult alighting on an upper leaf, is chronicled in the lower left and the right side of the print. Here, as in the majority of Merian's studies,

9 Maria Sibylla Merian (painter); Pieter Sluyter (engraver), *Poivre d'Inde*.

the insect is her primary concern. The moth is drawn and colored with great delicacy, while the pepper, although botanically correct, lacks a similar refinement. Her use of fine lines to delineate a form facilitated the task of the engraver, but resulted in a less naturalistic image than that achieved by later illustrators such as Pierre-Joseph Redouté.

1. Merian was the granddaughter of the Dutch engraver Johann Theodor de Bry (1562-1620); daughter of the Swiss engraver Matthäus Merian the Elder (active ca. 1624-41); stepdaughter of the Dutch flower painter Jacob Marrell (1614-81); and wife of Johann Graff (1637-1701), a student of Marrell. Wilfrid Blunt assisted by William T. Stearn, *The Art of Botanical Illustration* (London: Collins, 1950), p. 127. For a discussion of Merian's life and work see Evelyn G. Hutchinson, "The Influence of the New World on the Study of Natural History," in *The Changing Scenes in the Natural Sciences: 1776-1976*, ed. Clyde E. Goulden (Philadelphia: Academy of Natural Science, 1977).
2. Blunt, *Botanical Illustration*, p. 127.
3. Ibid. A partial bibliography of Merian's publications is given in Gordon Dunthorne, *Flower and Fruit of the 18th and Early 19th Centuries* (Washington, D. C.: Gordon Dunthorne, 1938), p. 221.
4. Merian's elder daughter, Johanna, later traveled to Surinam and provided her mother with additional drawings and specimens for a second edition of the *Metamorphosis*. Blunt, *Botanical Illustration*, p. 128; Dunthorne, *Flower and Fruit Prints*, pp. 21, 221.
5. These seventy-two plates were the same as in *Dissertation de Generatione et Metamorphosibus Insectorum Surinamensium*, Amsterdam, 1719; sixty of these had first appeared in the 1705 *Metamorphosis Insectorum Surinamensium*. Dunthorne, *Flower and Fruit Prints*, p. 221.

10 Pierre-Joseph Redouté (painter), French, 1759-1840; Chapuy (engraver), French. *Rosa Villosa, Pomifera*. Stipple engraving printed in colors and finished by hand, 13-1/2 x 10-1/8 inches (34.3 x 25.7 cm.), from Pierre-Joseph Redouté and Claude Antoine Thory, *Les Roses* (Paris: Imprimerie de Firmin didot, 1817-24). Dunthorne 232. Gift of The Print Club of Cleveland, In Honor of Mrs. William G. Mather. CMA 55.470

Pierre-Joseph Redouté has been called the portraitist of flowers. With sensitivity and intuition he captured more than their form and coloration; he revealed the essence of their ephemeral beauty, giving each bloom a distinct identity. The most celebrated flower painter of his day,[1] and indeed one of the most popular in the history of botanical art, Redouté is most famous for the illustrations in two of his many botanical works: *Les Liliacées*, an eight-volume folio with 486 plates published in Paris between 1802 and 1816; and *Les Roses*, a three-volume folio with 168 plates published in Paris, 1817-1824.[2] These delicately detailed studies, painted first on vellum with water color and then reproduced as stipple engravings,[3] are unexcelled flower prints, botanically accurate and possessing the quality of a fine color print.

In *Rosa Villosa, Pomifera* [10], Redouté faithfully records the life cycle of the apple rose as it develops from a tightly closed bud to a full-blown bloom.[4] A detail of the rose hip, or seed pod, is drawn to the right of the stem. The carefully controlled gradation of the colored inks reinforces the character of the life-sized image; the deep orange red of the hip offsets the delicate pink of the blossom. The variegated, softly hairy-textured, gray-green leaves have a crisp, naturalistic quality that effectively complements the texture and color of the softer flower and bristly hip.

Redouté, like Maria Merian, came from a family of fine artists. He was the son and grandson of established Belgian painters, and received his early training as a painter in his father's studio. Later he was greatly influenced by the paintings of Jan van Huysum (1682-1749), a Dutch painter in the baroque manner who created lush bouquets with trompe-l'oeil dewdrops and insects. While still a young man, Redouté made France his adopted home and studied with Gerard van Spaëndonck, a follower of van Huysum's. Redouté received his instruction in botany from Charles Louis L'Héritier de Brutelle, a wealthy botanist and staunch Linnean who opened his magnificent library to Redouté and instructed him in botanical disciplines.[5] The genius of Redouté as a scientific draftsman undoubtedly rests on this twofold training and on his full understanding of the structure and character of the plants he portrayed. Of equal importance is his ever-apparent appreciation for the pristine beauty of a single flower.

In the fall of 1828 Redouté was visited by John James Audubon, the American wildlife painter of growing eminence, who paid tribute to this aging "Raphael of Flowers":[6] "I had the pleasure of seeing old Redouté this morning, the flower painter *par excellence*. . . . His flowers are grouped with peculiar taste, well drawn and precise in the outlines, and coloured with a pure brilliancy that depicts nature incomparably better than I ever saw it before. Old Redouté dislikes all

that is not *nature alone*: he cannot bear the drawing of either stuffed birds or quadrupeds, and evinced a strong desire to see a work wherein nature was delineated in an animated manner."[7]

1. His popularity was such that, through the course of his career, he served as drawing master to Marie Antoinette, Queen of France; the Empresses Joséphine and Marie-Louise; the Duchess de Berry; and Queen Marie-Amélie. He was also a professor at the Musée Nationale. Sacheverell Sitwell and Roger Madol, *Album de Redouté* (London: Collins, 1954), p. 8.

2. A complete Redouté bibliography is given in Sitwell and Madol, *Album de Redouté*.

3. A stipple engraving, in which the design areas are made up of small dots or flecks, was produced by means of needles or various rollers on a single copperplate. Craftsmen applied the different colored inks to the copperplate with brushes or an engraver's dauber, and the inking process was repeated for each print made.

4. The apple rose, also known as Wolley-Dod's Rose, was introduced to Europe from western Asia in 1771. It is remarkable for its dark red, gooseberry-like fruit. Fred J. Chittenden, ed., *The Royal Horticultural Society and Dictionary of Gardening: A Practical and Scientific Encyclopedia of Horticulture* (Oxford: Oxford University Press, 1951), vol. 4, s.v. "R. Villosa." Gordon Edwards, *Wild and Old Garden Roses* (New York: Hafner Press, 1975), p. 42.

5. Wilfrid Blunt assisted by W. T. Stearn, *The Art of Botanical Illustration* (London: Collins, 1950), p. 174.

6. André Lawalrée and Günther Buchheim, *P. J. Redouté: Facsimile Prints Made from Mostly Unpublished Original Paintings by Pierre-Joseph Redouté* (Switzerland: Gesellschaft Sweizerischer Rosenfreunde; Pittsburgh: The Hunt Institute for Botanical Documentation, 1972), p. 14.

7. Sitwell and Madol, *Album de Redouté*, p. 14.

10 Pierre-Joseph Redouté (painter); Chapuy (engraver), *Rosa Villosa, Pomifera*.

11 Pierre-Joseph Redouté (painter), French, 1759-1840; Charlin (engraver), French. *Rosa Damascena, Celsiana.* Stipple engraving printed in colors and finished by hand, 13-7/8 x 10-1/8 inches (35.2 x 25.7 cm.), in Pierre-Joseph Redouté and Claude Antoine Thory, *Les Roses,* vol. 2 (Paris: Imprimerie de Firmin Didot, 1817-24). Dunthorne 232. The Garden Center of Greater Cleveland, Eleanor Squire Library, The Warren H. Corning Collection.

Les Roses, a three-volume work published by Redouté in Paris in 1817, marked the culmination of years of effort by the artist and Claude Antoine Thory (1759-1827), who wrote the folio's text. The roses were painted from life — Redouté set up his easel and canvas in the parks of Paris, Versailles, and Sèvres. He and Thory searched for new species in the fields and cultivated several varieties by planting, grafting, and multiplying their plants with cuttings.[1]

A small, special edition of *Les Roses* was printed with a portrait frontispiece and 169 plates in two states, uncolored and colored. Two original, colored drawings by Redouté were also included in these especially fine copies, which were bound in half-red morocco uncut leather. This volume is opened to the damask rose, *Rosa Damascena, Celsiana* [11], named after Cels, a French plant breeder who introduced it into France in 1750 from Holland.[2]

The engraving realistically portrays the character of the delicate flower with fine

11 Pierre-Joseph Redouté (painter); Charlin (engraver), *Rosa Damascena, Celsiana.*

25

lines and subtly shaded colors. The wide, mature bloom in the center of the bough is a very pale pink — almost white; it is framed by the younger, light pink and warm red flowers behind it, and each has a bright yellow stamen. The loose, softly folded petals of the flowers are complemented by the crisp, variegated, grayish-green foliage.

Redouté discussed his art and its importance in the preface to another of his folios: "The art of painting flowers is not a superfluous art, and the lavishness which lends itself to the ornamentation should not impair the true value of its usefulness. Natural history should not be deprived of its help.... How much these masterpieces have increased the number of friends of botany!"[3]

1. André Lawalrée and Günther Buchheim, *P. J. Redouté: Facsimile Prints Made from Mostly Unpublished Original Paintings by Pierre-Joseph Redouté* (Switzerland: Gesellschaft Sweizerischer Rosenfreunde; Pittsburgh: The Hunt Institute for Botanical Documentation, 1972), p. 10.
2. The frontispiece portrait of Redouté was engraved by C. S. Pradier after Gérard; Gordon Edwards, *Wild and Old Garden Roses* (New York: Hafner Press, 1975), pp. 71-72.
3. Lawalrée and Buchheim, *P. J. Redouté*, p. 13.

12 Philip Reinagle (painter);
Joseph Constantine Stadler (engraver),
Large Flowering Sensitive Plant.

12 Philip Reinagle (painter), English, 1749-1833; Joseph Constantine Stadler (engraver), English, active 1780-1812. *Large Flowering Sensitive Plant.* Aquatint and stipple engraving printed in colors and finished by hand, 14-1/16 x 17-1/2 inches (35.9 x 44.5 cm.), from Robert John Thornton, *The Temple of Flora* or *Garden of Nature. Picturesque Botanical Plates of the New Illustration of the Sexual System of Linnaeus* (London: Thornton, 1799-1807). Dunthorne 301. Gift of Robert Hays Gries. CMA 53.31

The Temple of Flora, one of the most famous of all florilegia, is the concluding portion of an immense, three-volume work by Robert John Thornton, MD, who titled himself a "public lecturer on medical botany." This work was announced to the public in 1797 and completed by 1807 as a *New Illustration of the Sexual System of Carolus van Linnaeus: Comprehending an Elucidation of the Several Parts of the Fructification; A Prize Dissertation on the Sexes of Plants; A Full Explanation of the Classes, and Orders, of the Sexual System; and the Temple of Flora, or Garden of Nature, Being Picturesque, Botanical, Coloured Plates, of Select Plants, Illustrative of the Same, with Descriptions.*

The publication of this book coincided with the rapid development of botany and the biological sciences at the turn of the century. Strange flowers from distant lands were being newly cultivated, lighting the imaginations of poets and leading to the publication of great botanical works with colored plates. Very much a product of its age, the text — much of it poetry — and the plates of the *New Illustration* combined elements of romanticism and neoclassicism with astute observations and notations. Despite the ambitions of its author, however, the book offers little of lasting scientific importance. It is, rather, a lavish glorification of botany and a tribute to the alliance of botany and the arts of painting and engraving — a tribute primarily evident in the thirty-one plates of *The Temple of Flora.*[1]

Thornton was assisted in his work by a team of professional painters and engravers, among them Philip Reinagle (1749-1833) and Joseph Constantine Stadler (active 1780-1812). The plates were executed by various processes — aquatint, mezzotint, stipple and line engraving — and some combined several of these methods. They were painted in basic colors and the proofs were then reworked by hand with watercolor washes. Thornton "directed" his flower pictures, choosing the plants, deciding upon symbolism, and arranging the backgrounds.[2] The plate *Large Flowering Sensitive Plant* [12], depicts a stately specimen standing with its crimson fronds and fan-shaped leaves open to the light, attracting native hummingbirds, while in the background an aborigine studies a second plant. The accompanying text describes the growth patterns, reproductive organs, and general attributes of the *Mimosa Grandiflora*, a native of the East and West Indies that was introduced into England in 1769. Although such information contributes to the study's botanical value, in general the *Large Flowering Sensitive Plant* remains a romantic narrative rather than a scientific inquiry.

1. Geoffrey Grigson and Handasyde Buchanan, *Thornton's Temple of Flora* (London: Collins, 1951), pp. 2, 3, 11.
2. Thorton drew the original for one of the rarest plates in the *Temple of Flora*, a group of roses. Reinagle, who was adept at portraits, sporting pieces, and landscapes, became a member of the Royal Academy. Grigson and Buchanan, *Thornton's Temple*, pp. 4, 10; see also Wilfrid Blunt assisted by W. T. Stearn, *The Art of Botanical Illustration* (London: Collins, 1950), pp. 203-8.

13 John James Audubon, American, 1785-1851. *Duck Hawks.* Oil on canvas, 1826-29, 25-1/2 x 36 inches (64.7 x 91.4 cm.). Gift of The American Foundation for the Maud E. and Warren H. Corning Collection. CMA 64.351

"I have *never* drawn from a stuffed specimen . . . nature *must* be seen first alive, and well studied before attempts are made at representing it."[1] Thus opposed to the prevailing method of painting birds and animals, Audubon based most of his vivid depictions on firsthand observations, and whenever possible painted directly from freshly caught specimens which he often wired in lifelike positions.[2] In doing so, he achieved more than a realistic record of form and plumage; he conveyed the essence of the living bird in action and in its natural habitat. Audubon was, for the most part, a self-taught artist, and he protested that he was not a "learned" naturalist but a practical one. Despite these limitations and his rudimentary understanding of the science of ornithology, Audubon resolved, about 1820, to draw all the birds of America and publish the results.[3] In search of specimens, he traveled from Labrador to regions as far south as Florida and Texas.

This project was realized in *The Birds of America*, a collection of 435 hand-colored, double-elephant folio-size (about 40 x 30 inches) engravings published from 1827 to 1838.[4] This volume established Audubon as the foremost nature painter in both Europe and America. Although some of the studies are uneven in technical control and scientific accuracy, Audubon's best images represent a brilliant fusion of precise observation and artistic skill.

Audubon is usually considered a watercolorist, but between 1826 and 1829 he repainted some of his most striking and popular compositions in oil in order to finance his publication. *Duck Hawks* [13], is based on plate 16 in *The Birds of America*, and depicts a male and a female peregrine falcon after killing a green-winged teal and a gadwall. The image combines a visual account of the birds' character with correct ornithological detail. Flying feathers, the bloodied beak of the female, and the defensive stance of the male dramatically illustrate the habits and temperament of the swift predators, while the spread wings and clawing talons display their coloration and physical forms.

In his journal, Audubon noted that the female falcon was collected on Christmas Day, 1820, during a journey down the Mississippi River, and was drawn in pastel shortly afterward. Later, probably after a hawk shoot near Niagara Falls in August 1824, he completed the original composition by combining the pastel with watercolor renderings of the male and the brace of ducks.[5]

1. Marshall B. Davidson, Introduction to *The Original Water-Color Paintings by John James Audubon: For the Birds of America*, vol. 1 (New York: American Heritage Publishing Co., 1966), p. xxi.
2. Edward H. Dwight, Introduction to *Audubon Watercolors and Drawings*, exh. cat. (Utica: Munson-Williams-Proctor Institute, 1965), p. 5.
3. Davidson, Introduction, p. xxviii.
4. Fewer than 200 double-elephant folios of *Birds of America* were originally published; 133 complete sets are known to survive. One of these is owned by and on display at The Cleveland Museum of Natural History. Waldeman H. Fries, *The Double Elephant Folio: The Story of Audubon's Birds of America* (Chicago: American Library Association, 1973).
5. *The Original Water-Color Paintings by John James Audubon: For the Birds of America*, vol. 1 (New York: American Heritage Publishing Co., 1966), pl. 315.

13 John James Audubon, *Duck Hawks.*

II Material Sciences

Materials — solid substances such as wood, stone, metal, and clay — are generally considered incidental to what is made from them. Throughout history, however, what man could make with these natural resources has been determined by his understanding of their properties.

Metal, for example, has the property of resisting stress only up to a certain point, and then it deforms, or bends, without breaking. It may be shaped by applying high stress, as by hammering, for example, and then it will resist the less severe stresses of use indefinitely. Man has made use of this metallic quality for millennia, but science has only recently revealed that this property of metal, and indeed the properties of all materials, result from the arrangement of atoms into masses, or aggregates, and the unique, complex distribution of electrons and protons within the atomic structure. The theory of deformation explains how and why metal bends, but scientific theory was not necessary for early craftsmen to discover that metals are deformable. It was enough to find out that metal exhibited certain properties that were reproducible, and then those qualities, which could be identified and put to use, were simply attributed to the unique nature of metals.[1]

The extent to which real, if untheoretical, knowledge was applied by early craftsmen working in glass, metal, or clay and using acids and pigments is impressive. They sensitively manipulated plastic and viscous flow, crystallization, surface tension differences, and color changes resulting from ions in various states of oxidation and polarization. They enjoyed the beauty imparted to a surface by chemical degradation and the irregularity that comes from fracture, deformation, and the sectioning of polycrystalline materials.[2] These qualities and effects were first appreciated and controlled by artists who were sensually aware of textures, colors, and the other physical properties of materials; today, the cause and nature of these properties can be explained by material scientists.

The study of material sciences is concerned with properties and the dependence of properties upon structure. That is, it examines and explains the solid-state physics of matter and the chemistry of its composition.[3] It has two primary goals: first, to determine and elucidate the physical and chemical properties of substances; and second, to develop and control chemical processes to attain specific results, The science of materials is a new field of study and its history is short. The histories of physics and chemistry, however, which are its major facets, are ancient, complex, and closely intertwined.[4]

The first theories about chemistry and the nature of matter came from the philosophers of China, India, and Greece — but many properties of metals and alloys had been discovered and put to use at least a millennium before. Primeval chemistry, which began with the kindling of fire, had made possible the transformation of soluble clay into insoluble pottery as well as the melting of sand with other substances to make glass as early as 4000-3500 BC.[5] These processes, however, were developed on the basis of practical, not theoretical, knowledge.

Often the artisan and the experimental scientist shared a heritage of experimentation. Sometimes the beauty and desirability of decorative objects actually stimulated scientific progress. At other times, the technical knowledge of artists and craftsmen was directly applicable to science. In the eighteenth century, for example, the European desire to duplicate Oriental porcelain inspired experiments in high-temperature chemistry and led to the first feasible chemical analysis for materials other than precious metals.[6]

Examples of the interplay between craftsman, artist, and scientist may be found throughout history. The use of corrosives may be traced as far back as 3000 BC to the Harrapa culture of Chanhadaro, India, where craftsmen used an alkaline substance and heat process to etch carnelian beads. In Central and South America, pre-Columbian Indians used acid corrosives to create a gold surface through a process now known as "depletion gilding." In Europe during the Middle Ages, craftsmen used a mixture of salts or a preparation of nitric acid to decoratively etch armor and weapons, and fine artists applied the technology to steel and copper printing plates. Recipes for these medieval etching agents were recorded in the craft manuals of the day, and this firsthand knowledge eventually contributed to the development of chemical theory and practice.[7]

1. Cyril Stanley Smith, "Matter versus Materials: A Historical View," *Science* 162 (November 1968): 638; idem, "Materials and the Development of Civilization and Science," ibid. 148 (14 May 1965): 910.
2. Smith, "Matter versus Materials," p. 638.
3. Solid-state physics involves the study of the physical properties and atomic structure of solids, particularly crystals. In addition to elasticity, magnetism, and specific heat, it is also concerned with the relationship between the purity of a metal and its strength and ability to conduct electricity.
4. For a good synopsis of the development of the science of materials see Smith, "Matter versus Materials," pp. 637-44; idem, "Materials and the Development," pp. 908-17; idem, "Art, Technology, and Science: Notes on Their Historical Interaction," in *Perspectives in the History of Science and Technology*, ed. Duane H. D. Roller, (Norman: University of Oklahoma Press, 1971), pp. 129-65. For a general history of science and the humanities see Jacob Bronowski, *The Ascent of Man* (Boston: Little, Brown and Co., 1973).
5. Fritz Ferchl and A. Süssenguth, *A Pictorial History of Chemistry* (London: William Heinemann, 1939), p. 6; Smith, "Matter versus Materials," p. 638.
6. Smith, "Art, Technology, and Science," p. 137; idem, *Aspects of Art and Science*, exh. cat. (Cambridge: Margaret Hutchinson Compton Gallery, Massachusetts Institute of Technology, 1978).
7. Jon Eklund, "Art Opens Way for Science," *Chemical & Engineering News* (5 June 1978): 31; Smith, *Aspects of Art and Science*, pp. 29-34.

14 North American, Navajo Tribe. *Indian Dye Chart*. Various plant materials and wool, ca. 1973/74, 21-3/4 x 27-3/4 inches (55.2 x 70.5 cm.). The Harold T. Clark Educational Extension Fund. CMA 74.1060

Dyes are complex chemical compounds that fall into various classifications, depending on their molecular structure. Basically, they are soluble in water and transfer their color, often chemically, to the materials to which they are applied.[1] Each natural dye, made from either animal, vegetable, mineral, or metal dyestuffs, has a unique chemical composition, and each reacts differently with various fibers. Mordants, which include among other substances alum, urine, vinegar, and copper sulfate, have been used since ancient times to change the color of a dye or make it more permanent — the result of a chemical reaction between the mordant and dye which causes an insoluble salt to form that coats the fibers being colored, and allows the colorant to bite into them.[2] It was the need for mordants, and especially alum, that gave rise to the first large-scale chemical industry.

Man learned early that color enhances the appearance of his possessions, and achieved a wide range of colors through experimentation with available resources. The ancient Peruvians, for example, developed almost 200 different hues.[3] This *Indian Dye Chart* [14], displays some of the flowers, barks, twigs, roots, and berries used for dyes by the Navajo Indians of the American Southwest. The top row, from left to right, includes: red gila (pink); sagebrush (yellow green); juniper berries (brown); gambel oak bark (tan); snakeweed (yellow); Indian paint brush

31

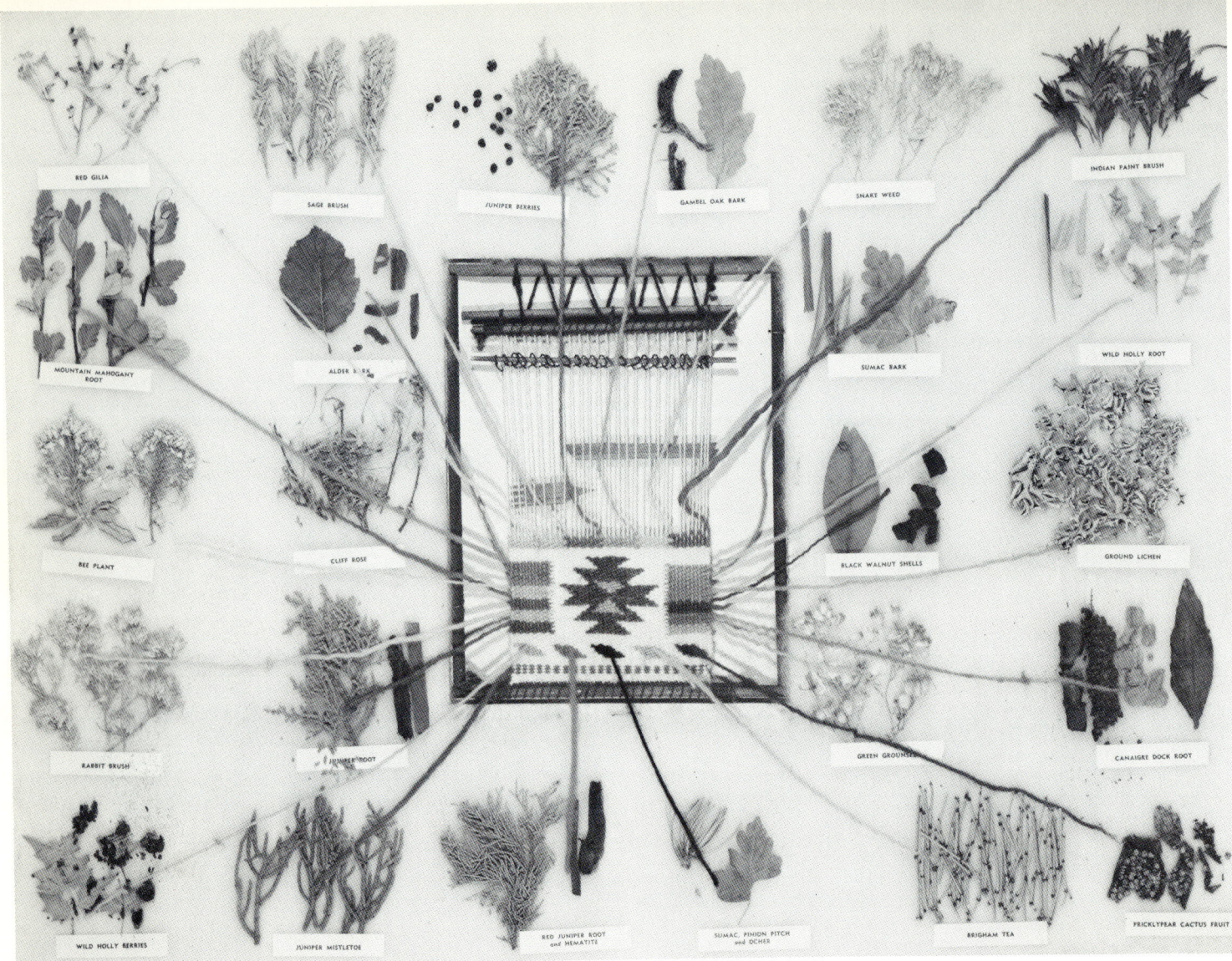

(dark brown). The second row: mountain mahogany root (deep pink); alder bark (light brown); sumac bark (light brown); wild holly root (light green). The third row: bee plant (ivory); cliff rose (dark brown); black walnut shells (dark brown); ground lichen (orange). The fourth row: rabbit brush (dark yellow); juniper root (dark brown); green groundsel (yellow green); canaigre dock root (light brown). The fifth row: wild holly berries (light purple); juniper mistletoe (dark green); red juniper root and hematite (red brown); sumac, pinion pitch, and ocher (black); brigham tea (gold); prickly pear cactus fruit (dark purple). These rich, subtle earth tones were obtained by crushing the plant material, steeping it in water and perhaps a mordant for a period of time, and then straining the mixture.

1. In addition to textiles, dyes were used to color wood, bone, ivory, leather, plant fibers, and later, stones. Since they were absorbed by the material, dyes were often more permanent than paint. Henry Hodges, *Artifacts: An Introduction to Early Materials and Technology* (London: John Baker, 1964), pp. 160-61.
2. Fritz Ferchl and A. Süssenguth, *A Pictorial History of Chemistry* (London: William Heinemann, 1939), p. 8; Verla Birrell, *The Textile Arts: A Handbook of Weaving, Braiding, Printing, and Other Textile Techniques* (New York: Schocken Books, 1976), pp. 389-400; Cyril Stanley Smith, *Aspects of Art and Science*, exh. cat. (Cambridge: Margaret Hutchinson Compton Gallery, Massachusetts Institute of Technology, 1978), pp. 35-37.
3. Birrell, *Textile Arts*, pp. 379, 395. The range of colors available from natural dyes could be extended by over-dyeing. Hodges, *Artifacts*, p. 161.

14 *Indian Dye Chart.*

15 Egypt, Byzantine Period (AD 395-640). *Fragment of Ornamental Band from a Tunic or Cloth*. Wool, linen, and natural dyes, 5-1/2 x 20-1/2 inches (14 x 52.1 cm.). Gift of George D. Pratt. CMA 29.98

In Egypt the art and chemistry of dyeing were known as early as the predynastic period (ca. 5000-ca. 3000 BC). By 2000 BC the dyeing of linen and leather was both an individual craft and a temple industry, and during the Ptolemaic period (305-30 BC) it became a state monopoly, although private dyeing was permitted with the purchase of a dyeing license.[1]

When the *Fragment of Ornamental Band from a Tunic or Cloth* [15], was woven in

15 *Fragment of Ornamental Band from a Tunic or Cloth.*

33

the sixth century the Egyptian dye industry was quite sophisticated, and craftsmen could refer to recipe books describing the use of materials such as alkanet, safflower, saffron, kermes, madder, and woad.[2] The characteristic colors of several of these dyestuffs are found in the geometric designs and portrait medallion of this fragment.

The deep red of the background — kermes — was made from the brilliant scarlet scales of the female kermes insect from the Mediterranean region. Preparation of this dye simply involved collecting, drying, and crushing the insects.[3] The blue of the border framing the square motifs — indigo — was extracted from the indigo plant. The light green within the square motifs and the woman's tunic resulted from a combination of indigo and nut galls, while the thin blue-green stripes bordering the strip were colored by first dipping the fibers in an oak gall dye and then in indigo. The tan or buff color of the woman's face and other areas of the design was perhaps derived from fustic, a common dyewood.

1. R. J. Forbes, "Chemical, Culinary, and Cosmetic Arts," in *A History of Technology*, eds. Charles Singer, E. J. Holmyard, and A. R. Hall (Oxford: Clarendon Press, 1955), vol. 1, *From Early Times to Fall of Ancient Empires*, p. 248.

2. Ibid., p. 249.

3. Kermes (*coccus ilicis*) contains the coloring matter kermesic acid, $C_{18}H_{12}O_9$, and was probably used with an alum mordant. The word "kermes" is Arabic in origin and the source of the English word "crimson." Similarly, the English word "vermilion" comes from the Latin *vermiculum*, or little worm. Rutherford John Gettens and George L. Stout, *Painting Materials: A Short Encyclopaedia* (New York: D. Van Nostrand Co., 1942), p. 123.

16 France, Sain-Bel. *Blue Indigo Resist Print*. Linen and indigo, 1790, 20 x 20-1/2 inches (50.8 x 52.1 cm.). Gift of Gertrude Underhill. CMA 35.113

Unfortunately, most of the ancient, natural dyes produced unstable, fugitive colors that soon faded when exposed to sunlight. There were notable exceptions, however, such as the reds obtained from the madder root and the lac insect, or the blue of the indigo plant. Indigo, the dye used to make the *Blue Indigo Resist Print* [16], is a deep blue coloring matter developed very early in the Far East for dyeing and painting. Its use as a dye is mentioned in an Egyptian papyrus from about 2000 BC, and it has been identified as one of the pigments used for decorating Roman parade shields around AD 200. It appears in European commercial transactions as early as the thirteenth century. As with all natural dyes, the effectiveness of indigo rests on its unique physical properties and on their chemical reactions to other organic compounds.[1]

The indigo plant, *indigofera*, is a member of the pea family and, as its name implies, probably originated in India. Indigo exists in the plant as a colorless glycoside called "indican." It is freed from the plant by a fermentation process, and oxidizes to blue when exposed to air. Indigo was produced in many parts of the world. In the late thirteenth century Marco Polo described the Chinese process: "Indigo also of an excellent quality, and in large quantities, is made here. They procure it from an herbaceous plant, which is taken up by the roots and put into tubs of water, where it is suffered to remain until it rots, when they press out the juice; this on being exposed to the sun and evaporated, leaves a kind of paste which is cut into small pieces of the form in which we see it brought to us." For use with textiles the deep violet-blue crystallized dye must be deoxidized, or reduced, and returned to a liquid form. This is accomplished by dissolving the indigo in a reducing agent such as urine. The fibers absorb this colorless, organic, soluble compound. As these fibers dry, the dye is then re-oxidized by the air to form the insoluble indigo blue.[2]

Indigo was imported to Europe as early as 1516, but not introduced generally until about 1602. Since indigo is ten times stronger than woad, its rival European dye, its use was strongly opposed by the woad growers, and prohibitive laws were enacted both on the Continent and in England. In 1737, however, indigo was legally permitted in France, and its valuable properties were gradually recognized throughout Europe. The *Blue Indigo Resist Print*, made in France about fifty years later, offsets the white of the natural linen against the rich indigo blue in the rhythmic repetition of a formal leaf and curved stem. This silhouetted design, which is reminiscent of traditional Asian motifs, was made by impregnating the design part of the fabric with hot wax, thereby blocking its absorption of dye.[3]

16 *Blue Indigo Resist Print*.

1. Rutherford John Gettens and George L. Stout, *Painting Materials: A Short Encyclopaedia* (New York: D. Van Nostrand Co., 1942), p. 120.

2. During the fermentation process the glycoside is hydrolized to indigo and sugar; the coloring matter is indigotin, $C_{16}H_{10}N_2O_2$. Gettens and Stout, *Painting Materials*, p. 120; Verla Birrell, *The Textile Arts: A Handbook of Weaving, Braiding, Printing, and Other Textile Techniques* (New York: Schocken Books, 1976), p. 385.

3. Gettens and Stout, *Painting Materials*, p. 121; Birrell, *Textile Arts*, p. 385. The design was printed in wax by means of a wood block.

17 Egypt, 18th Dynasty (1580-1314 BC), Reign of Amenophis II. *Paint Box of Amenemopet*. Boxwood, graphite, Egyptian green, Egyptian blue, and ocher, L. 8-1/4 inches (21 cm.). Gift of The John Huntington Art and Polytechnic Trust. CMA 14.680

The *Paint Box of Amenemopet* [17] is one of the many archeological discoveries that document the pigments used by early civilizations. At first glance it appears to be a funeral object that was buried with the deceased for use in the next life. In fact, however, the box was used considerably during the earthly life. The first and second cakes of black are smeared, as is the light blue one next to them. The heavy depressions, especially in the black cakes, suggest that the box was used by Amenemopet as a writing palette. The cakes of color were probably made by mixing finely ground pigments with gum and water and letting them dry. As with modern water colors, they would have been used by dipping a brush in water and rubbing it on the ink.[1]

Pigments differ from dyes both in their chemical composition and physical qualities. Pigment molecules are larger than those of dyes, and are insoluble in water. They have no adhesive qualities, and so, to be useful as paints or inks, their discrete particles must be suspended in a vehicle or medium. The color characteristics, the hue, purity, and brightness of the light diffused by a particular pigment depend upon the color absorption, size, shape, and texture of the pigment grains.[2]

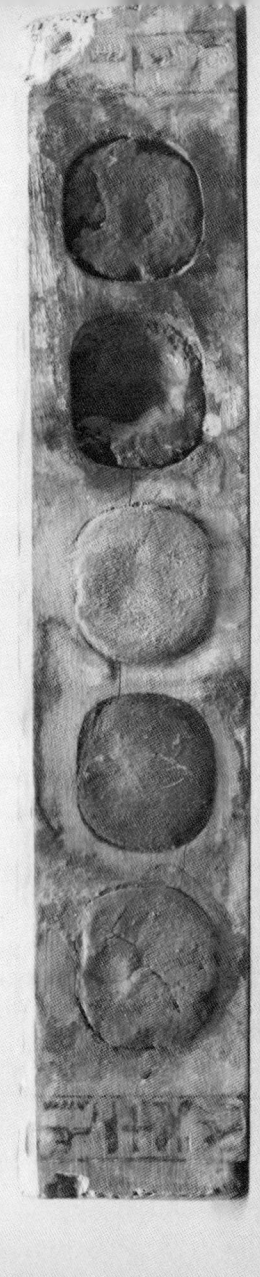

Pigments are made up of many different chemical compounds with varied chemical properties. Throughout history they have been derived from a variety of substances, organic and inorganic, natural and artificial. Carbon black in the form of soot, charcoal, or even charred bones was a natural by-product of fire and one of the first pigment materials known. Graphite, another black pigment, is a soft, black or dark gray substance widely existing in nature. One of the principal Egyptian sources for graphite is in certain schists in gold-mining areas of the eastern desert. The natural earth pigments — ochers, umbers, and siennas — were obtained from earthy deposits containing ferric oxide as a colorant. White could be made by crushing several naturally occurring deposits, such as chalk, lime, gypsum, and china clay.

Both natural and artificial pigments are contained in the paint box. The two black inks are graphite, and the red at the opposite end of the box is a naturally occurring ocher. The light blue and pale green at the center of the box are artificial pigments.

17 *Paint Box of Amenemopet.*

Egyptian blue, the principal blue pigment of ancient Egypt, consisted of a crystalline compound of silica, copper and calcium made by heating together silica, a copper compound (probably malachite), calcium carbonate, and natron. Egyptian green was based on a similar formula.

1. The Cleveland Museum of Art, Ancient Department, Curator's file; A. Lucas, *Ancient Egyptian Materials and Industries*, 4th ed., rev. and enl. by J. R. Harris (London: Edward Arnold, 1962), pp. 340-45, 362-63.
2. Rutherford John Gettens and George L. Stout, *Painting Materials: A Short Encyclopaedia* (New York: D. Van Nostrand Co., 1942), pp. 138, 143-45; Henry Hodges, *Artifacts: An Introduction to Early Materials and Technology* (London: John Baker, 1964), pp. 156-57.
3. Hodges, *Artifacts*, pp. 156-57.
4. Lucas, *Ancient Egyptian Materials*, pp. 340-41, 348.

18 Italy, 13th century. *Nine Historiated Initials*, probably from a Bible. Tempera and gold leaf on vellum, H. of largest initial: 4-3/8 inches (11.2 cm.). Gift of Vittorio Forti. CMA 54.131-.139

19 France, Paris, Atelier of Boucicaut Master. *Risen Christ as Ruler of the World*, leaf from *Book of Hours*. Tempera and gold leaf on vellum, ca. 1410-15, 6-11/16 x 4-15/16 inches (17 x 12.5 cm.). Purchase from the J. H. Wade Fund. CMA 53.367

Modern organic chemistry is based, in part, on the accumulated knowledge of many generations of artists who searched for dyes and pigments. The illuminators' use of rich and luminous colors in the *Historiated Initials* [18] as well as in the leaf with the *Risen Christ as Ruler of the World* [19] not only exemplifies the skill and understanding of medieval craftsmen but also foreshadows later scientific advances.

The medieval painter needed more than an artistic inclination; success in his profession demanded the expertise of a geologist and a chemist. In most cases, the painter himself located the necessary ingredients for his work and produced his own pigments. Vermilion, for example, was produced from the reaction of mercury and sulfur. The most admired blue color was ultramarine, extracted from powdered lapis lazuli by a flotation process that was frequently described in manuals of the time. One of the best recipes was given by Cennino Cennini, a fourteenth-century artist who wrote *The Book of the Art*, a comprehensive account of techniques of that period. He directs the powdered mineral to be kneaded in a weak lye solution with a paste or dough of wax, pine resin, linseed oil, and gum mastic. The dough retains the foreign particles (among them silica, calcite, and pyrite), but the fine particles of blue color settle out in the alkaline water. The first extraction gives the finest and purest color; each successive extraction produces a less pure product. The separation is caused by the preferential wetting and retention of the impurities by the dough mixture, a process described today as the exploitation of differential interface energies.[1]

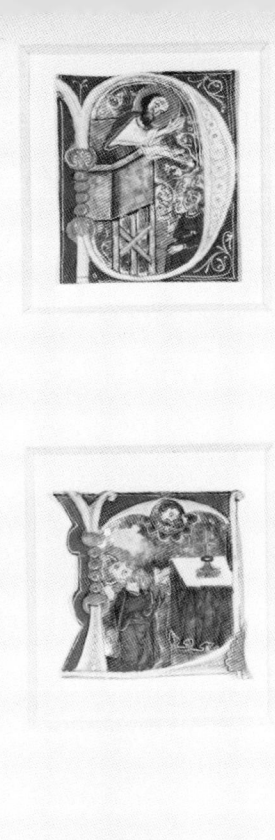

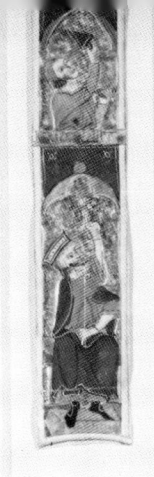

18 *Nine Historiated Initials.*

An important consequence of man's continued exploration of color was the emergence in the mid-nineteenth century of the branch of physics called spectrum analysis, which provided a new method of investigating the chemical nature of substances and led to the discovery of a number of new elements — and to reveal the structure of atoms and molecules.

1. Rutherford John Gettens and George L. Stout, *Painting Materials: A Short Encyclopaedia* (New York: D. Van Nostrand Co., 1942), pp. 165-66, 170-73; Christiana H. Herringham, *The Book of the Art of Cennino Cennini* (London: George Allen and Unwin, 1922), pp. 47-51.

2. Cyril Stanley Smith, *Aspects of Art and Science*, exh. cat. (Cambridge: Margaret Hutchinson Compton Gallery, Massachusetts Institute of Technology, 1978), p. 35.

Below left
19 *Risen Christ as Ruler of the World.*

Below
20 *Deer Effigy.*

20 Panama, Cocle, 14th-15th century. *Deer Effigy*. Tumbaga, L. 3-1/2 inches (8.9 cm.). Gift of Mrs. R. Henry Norweb. CMA 52.325

Man's practical understanding of acids and corrosive alkalies may be traced directly to the chemical technologies associated with art. The earliest-known decorative use of localized chemical attack was made more than 4000 years ago in India. There, artists in the Harrapa culture used an alkaline paint, which they subjected to heat treatment, to bleach abstract patterns into carnelian beads without removing much mass. In North America, from about 900 to 1200, Indians of the Snaketown Pueblo etched deep designs in seashells by using natural fruit juice and a resist similar to pine resin. The shell, with its high calcium carbonate composition, was easily attacked by the weak acid.[1]

The typical mineral acid, such as sulfuric or hydrochloric acid, reacts violently upon contact with many substances and has a destructive, disintegrating effect on their properties. An alkali, such as caustic soda or lye, is the exact opposite of acid, but will also react destructively with many substances, especially fats, oils, or waxes. Chemically, an acid is any substance capable of liberating hydrogen ions (an ion is an electrically unbalanced or charged atom). An alkali is one which yields hydroxyl ions.[2]

In Central and South America pre-Columbian Indians used acid corrosives to create a gold surface by means of a process known as depletion gilding. The small *Deer Effigy* [20], recovered from the Well of Sacrifice at Chichen Itza in Yucatan, was formed by this method.[3] Its curving body and stylized features were molded in an alloy known as tumbaga (usually comprised of about 25 percent gold and 75 percent copper), and treated with a natural, crude acidic substance that dissolved the baser metals from the surface of the effigy, leaving a thin layer of the precious gold.

Mineral copiapite, which contains a form of basic ferric sulfate hydrate that results from the natural weathering of pyrite in dry regions, was probably used for this treatment. It was also probably the substance described in 1580 by the Spanish priest Sahagun, in his account of Aztec crafts: tumbaga was finished, he said, by rubbing it with "what was called 'gold medicine.' It was just like yellow earth mixed with a little salt; with this gold was perfected and became very yellow. And later it was polished . . . so that at last it glistened, it shone, it sent forth rays." In Europe the ferric sulfates were known as *Bergbutter*, or misy, and formed the basis for most of the early corrosive mixtures used in the arts.[4]

1. Cyril Stanley Smith, *Aspects of Art and Science*, exh. cat. (Cambridge: Margaret Hutchinson Compton Gallery, Massachusetts Institute of Technology, 1978), pp. 30, 32; see also Henry Hodges, *Artifacts: An Introduction to Early Materials and Technology* (London: John Baker, 1964), p. 169.
2. Ralph Mayer, *The Artist's Handbook of Materials and Techniques* (New York: The Viking Press, 1977), p. 441.
3. It is one of group of "curly tailed animals," characteristic of Cocle ornaments.
4. Smith, *Aspects of Art and Science*, p. 32.

21 Italy, Milan, late 16th century. *Cabasset*. Etched steel, gilt and brass-headed rivets, L. 10-5/8 inches (27 cm.). Gift of Mr. and Mrs. John L. Severance. CMA 16.1525

The use of acids as a pickling solution for metals, to remove scale formed by annealing, and to blanch silver-bearing coins, is very old. The earliest written European reference concerning etching, however, dates from the eighth century and refers to preparing an iron surface for gilding. The first mention of an etching reagent that dissolved unoxidized metals and the earliest extant examples of decoratively etched armor date from the fifteenth century. A manuscript from 1409, which includes a note indicating that it had been transcribed from an earlier one, provides the following recipe:

> To make a water that corrodes iron — Take one ounce of sal ammoniac, one ounce of roche alum, one ounce of sublimed silver, and one ounce of Roman vitriol. Pound them well, take a glazed earthen vase, pour into it equal parts of vinegar and water, then throw in the above-mentioned articles. Boil the whole until reduced to half a cup or a cup; apply it to such parts of the iron as you may wish to hollow or corrode and the water will corrode them.[1]

Other corrosive mixtures of the period included such substances as verdigris, salt, alum, tartar, ferric sulfate, saltpeter, ammonium chloride, and mercury sublimate. Most were used as a paste. The etching process was a slow one; the mixture was left on for two or three days.[2]

Protective armor, such as this cabasset [21] worn by a sixteenth-century knight, could be decoratively etched in either of two ways. One was to cover the entire surface with wax or a linseed-oil paint, scratch the design through the thin coating, and then apply acid to fix this design in the metal. A second way was to draw the design with the protective substance, then etch away the background with corrosives.[3] Both methods were used to decorate this cabasset. The wide stripes of tiny representations of armor pieces that divide the crown into four triangular sections and the draped figure holding a lance and shield within the oval medallion in each section were made by the second method. The simpler, linear patterns within the sections were probably scratched through the protective layer and then etched. The gilding that originally heightened the contrast between the etched areas and the surrounding polished surfaces is now only barely visible.

Although its purpose was utilitarian, armor might be considered a kind of hollow sculpture based on the human form. Its design and construction required a practical understanding of human mechanics, an applied knowledge of metals and corrosives, and the development of ingenious engineering devices.

1. Cyril Stanley Smith, *A History of Metallography: The Development of Ideas on the Structure of Metals Before 1890* (Chicago: University of Chicago Press, 1960), p. 10; idem, *Aspects of Art and Science*, exh. cat. (Cambridge: Margaret Hutchinson Compton Gallery, Massachusetts Institute of Technology, 1978), p. 31.
2. Smith, *A History of Metallography*, idem, pp. 10-11; *Aspects of Art and Science*, p. 31.
3. Howard L. Blackmore, *Arms and Armour* (New York: E. P. Dutton and Co., 1965), p. 91; see also Charles Henry Ashdown, *European Arms and Armour* (New York: Brussel & Brussel, 1967).

21 *Cabasset.*

22 Czechoslovakia, Karlsbad, Moser Glassworks. *Rose Bowl.* Acid-embossed glass, 1939, D. 8 inches (20.3 cm.). Gift of Leonard C. Hanna, Jr., for the Coralie Walker Hanna Memorial Collection. CMA 39.194

22 *Rose Bowl.*

The etching technique used to create the underwater scene on this *Rose Bowl* [22] is said to have been developed by the German engraver Heinrich Schwanhardt about 1670. Known as acid embossing, it involves the use of hydrofluoric acid, the only known acid that will actively attack glass. As with metal etching, the glass surface is coated with a thin layer of acid-resisting compound, such as gum, wax, or varnish. The design is drawn through the coating with a fine steel point and the acid allowed to attack the exposed area. After a short time the glass is washed and the resist removed. With this bowl the hydrofluoric acid was used to remove the background of the design, leaving the motif in relief, its original clear surface contrasting with the etched, dull ground.[1]

The high-relief motif on the *Rose Bowl* consists of several fanciful fish of unidentifiable species, swimming among flowering plants and floating air bubbles. Linear details, such as the fishes' fins and body markings, were also etched into the glass surface. Their rough, bitten edges contrast sharply with the soft, ripple, sandpaper-like texture of the recessed areas and the original, smooth surface of the transparent, bluish-purple glass.

1. E. M. Elville, *The Collector's Dictionary of Glass* (London: Country Life, 1961), pp. 92, 105; Edward Dillon, *Glass* (London: Metheun and Co., 1907), pp. 281-82.

41

23 John Taylor Arms, American, 1887-1953. *Cancelled Plate for an Umbrian Street.* Etched copperplate that had been electrotyped, 1925, 10-3/16 x 5-5/16 inches (26 x 13.6 cm.). Gift of the Artist. CMA 26.197

24 John Taylor Arms. *An Umbrian Street.* Etching, 1925, 10-3/16 x 5-5/16 inches (26 x 13.6 cm.). Gift of the Artist. CMA 26.198

Both the engraved and etched plates used by graphic artists for printing came out of the metalworker's shop. Engraving, which considerably predated the use of chemical etching as a means of making recessed lines in a metal plate, grew out of the metallurgical technique of niello decoration. In this method, lines engraved with a burin were filled with a mixture of copper, silver, and lead sulfides. Ink was used to test the engraving before fusion, a practice that suggested the transfer of an engraved image to paper.

The etching technique used to decorate a piece of armor was very similar to that used to make an etched printing plate, and, apparently, the first etchings were printed from iron plates that had been prepared by armorers. The earliest-known etchings are believed to have been executed in 1503 although 1513 is the first verifiable date. In 1515 Albrecht Dürer, who had designed and decorated armor, began making prints that combined both chemical attack and engraving methods.[1]

23 John Taylor Arms,
Cancelled Plate for An Umbrian Street.

The etching agents for printing plates were not unlike those used by armorers. One seventeenth-century recipe combined three pints of vinegar and six ounces of both sal ammoniac and common salt, with four ounces of verdigris. These were boiled in a glazed pot and allowed to settle before use. Ferric chloride was later used in the etching process, and shortly after 1850 a Dutch mordant consisting of hydrochloric acid and potassium chlorate was introduced. The techniques used to etch the plate for the print *An Umbrian Street* [24] by John Taylor Arms were essentially the same as those used by medieval armorers and early printmakers.[2]

1. Cyril Stanley Smith, *A History of Metallography: The Development of Ideas on the Structure of Metals before 1890* (Chicago: University of Chicago Press, 1960), p. 11.
2. Ibid.

24 John Taylor Arms, *An Umbrian Street*.

25 Syria (?), AD 1st-11th century. *Glass Vase*. Blue glass, H. 2-3/4 inches (7 cm.). Gift of Mr. and Mrs. W. A. C. Miller, III. CMA 53.37

26 Asia Minor, AD 3rd-4th century. *Bottle*. Yellow glass, slightly iridescent, with applied decoration, H. 4-1/8 inches (10.5 cm.). Gift of The John Huntington Art and Polytechnic Trust. CMA 15.548

Glassmaking is a complex process in which dry, opaque materials are combined into a molten substance and then transformed into a hard, brittle, usually transparent object. There is today a steadily expanding range of scientific and technological applications for glass, whose thermodynamics and mechanical properties have recently been deciphered. These applications, however, will involve the same optical and thread-forming qualities of glass that artisans have depended upon for more than 5000 years.[1]

Glass is a supercooled liquid made by the fusion of soda, lime, and silica in varying proportions. Other ingredients, such as lead, barium, potassium, calcium, nitrate, arsenic, manganese, and iron may also be part of its composition; several metallic oxides may be added as colorants. However, since hundreds of different chemical combinations can result in glass, its chemical composition is not regarded as the fundamental factor which makes glass "glass." Today's scientists define glass, rather, as a substance in a glassy state, that is, a state in which the molecular units have a disordered arrangement, but sufficient cohesion to produce overall mechanical rigidity. The term "vitreous state" is frequently used to describe this condition. Over the centuries glass has often been worked in such a way as to imitate other materials, so that it has resembled metal, porcelain, or precious stone. Glass is perhaps most appealing, however, when its unique nature and peculiar qualities are readily apparent.[2]

The oldest forms of glassware known are the translucent pieces used for inlay work and colored beads made by the Egyptians and Babylonians during the third millennium BC. The uses of glass expanded quickly, however, as its chemical and physical nature became understood and controllable. The first glass vessels were made by wrapping molten glass around a heat-resistant core made of limelike material to which the glass would not adhere, and which, after cooling, was removed. Later, by the first century BC, craftsmen began blowing glass at the end of a long, hollow iron rod, first shaping the glass by blowing it into molds, but later blowing it free form.[3]

The mechanics of the glassblowing process and the nature of the material itself influenced the shape of glass vessels. The basic shape of all blown glass is a sphere.

25 *Glass Vase.*

26 *Bottle.*

Suspended on the tip of the blowpipe, the hot, pliable sphere is transformed, either by its own weight or by a swinging movement, into a club or tear shape. The spherical *Glass Vase* [25], thought to be Syrian, and the tear-shaped *Bottle* [26] from Asia Minor exemplify these basic shapes. After their initial formation, the bottoms of both were flattened. A slight indentation was made in the bottom of the vase, while a ring-shaped foot was added to the base of the bottle to improve stability. The shaped vessels were then cooled gradually to insure uniform contraction throughout the body of the glass. This process minimized porosity and gave strength and durability to the glass. After cooling, the neck was broken from the blowpipe and then, after reheating, enlarged into a funnel-like shape.[4] The mouth of the vase was folded over on itself to form a smooth, slightly raised lip.

Thickish glass threads that retain the character of molten glass are the only adornments. They were melted onto the surface of the bottle in three H-shaped patterns on the body of the vase and a ridged band at its shoulder. While primarily decorations, they also may have made the vessel easier to grasp. The glass threads attached to the vase however, are obviously handles, jutting out from the lip and then falling at a steep angle to the vase's shoulder. The elasticity of glass in its molten state is frozen in the graceful movement of the handles from the overlapping joint at the neck, through the slightly concave stem, to the bulbous shoulder attachments. The bottle's light green color is probably due to the unintentional presence of a copper compound or iron oxide in the glass body. The deep, rich blue of the vase, on the other hand, was intentionally produced, probably by adding cobalt to the mixture.[5] The white discoloration and slight iridescence of both vessels were caused by natural deterioration of the glass.

1. Cyril Stanley Smith, *Aspects of Art and Science*, exh. cat. (Cambridge: Margaret Hutchinson Compton Gallery, Massachusetts Institute of Technology, 1978), p. 23.
2. Some 75,000 different chemical compositions for glass are on file with the Corning Glass Works (Corning, New York) alone. For a detailed discussion of the glassy state of matter and its properties, see Robert H. Brill, "A Note on the Scientist's Definition of Glass," *Journal of Glass Studies* 4 (1962): 127-38; Rutherford John Gettens and George L. Stout, *Painting Materials: A Short Encyclopaedia* (New York: D. Van Nostrand Co., 1942), pp. 233-34; Fritz Kämpfer and Klaus G. Beyer, *Glass: A World History, the Story of 4000 Years of Fine Glass-Making*, trans. and rev. Edmund Launert (Greenwich, Conn.: New York Graphic Society, 1966), p. 9.
3. Kämpfer and Beyer, *Glass*, pp. 9-10.
4. Ibid., p. 32.
5. A. Lucas, *Ancient Egyptian Materials and Industries*, 4th ed., rev. and enl. J. R. Harris, (London: Edward Arnold, 1962), pp. 188-89.

27 Belgium (Flemish), 16th century. *Christ in the Garden of Gethsemane*. Stained and painted glass and lead, 29 x 28-1/4 inches (73.7 x 71.8 cm.). Gift of S. Livingston Mather, Constance Mather Bishop, Philip R. Mather, Katherine Hoyt Cross, and Katherine Mather McLean in accordance with the wishes of Samuel Mather. CMA 40.340

Glassworkers learned early that the addition of certain minerals to molten glass would produce certain colors, but the art and technology of stained and painted glass reached their highest level in the medieval glass window. New colors and working procedures were developed throughout the Middle Ages as craftsmen deepened their understanding of the chemistry and physics of glass. The sixteenth-century glassworker's practical knowledge of glass, metals, fluxes, and firing conditions and the skill with which he controlled the many structural changes that occurred during the glassmaking process are evident in the complex, small window, *Christ in the Garden of Gethsemane* [27]. Almost 150 pieces of glass in different shades and hues, and with enameled details and textures, compose the somber garden scene.

The solid-colored pieces of glass making up the red of the soldier's uniform, the blue of the sky, and the green areas around the seated disciples were stained by adding metallic oxides such as cobalt, copper, manganese, or iron to the mixture during melting. This produced an even tint of color not only on the surface but also through the entire piece of glass. In another method of staining, developed by craftsmen in the fourteenth century, silver sulfide was applied to the surface of the glass and heated. During the heating, the glass actually absorbed the silver-oxide colorant and acquired a yellow stain.[1]

The facial features of Christ, the disciples, and the soldiers; the drapery folds and landscape details; and the wood-graining on the fence in the foreground, as well as the ar-

27 *Christ in the Garden of Gethsemane.*

chitectural elements at the right of the window, were drawn with a mixture composed of metallic oxides and ground glass, perhaps suspended in a medium of gum arabic or oil. This enamel was painted onto the surface of clear or colored glass and heated to a high temperature until it became a thin layer of glass fused with the original piece.

1. Henry Hodges, *Artifacts: An Introduction to Early Materials and Technology* (London: John Baker, 1964), p. 62; see also Jaroslar R. Vavra, *5000 Years of Glass-Making: The History of Glass* (Prague: Artia, 1956), pp. 76, 81; Robert Sowers, *The Lost Art: A Survey of One Thousand Years of Stained Glass* (London: Lund Humphries, 1955), p. 14.

28 Bohemia, mid-19th century. *Covered Cup.* Ruby glass and clear glass, H. 23-1/8 inches (58.7 cm.). Gift of Mrs. S. Prentiss Baldwin. CMA 47.274

Throughout the Middle Ages, and even as late as the seventeenth century, the quest of the alchemist and the art of the glassmaker were closely related. One of their concerns was the formula for ruby glass, the material used to make the large *Covered Cup* [28]. Neither alchemists nor glassmakers were particularly interested in achieving the deep red color of glass for its own sake, but both believed that drinking from such glass would provide protection from many ills. Johann Kunckel, a seventeenth-century German alchemist and glass technologist, is widely

28 *Covered Cup.*

recognized — although some experts consider this a matter of debate — as the inventor of the first satisfactory process for making the long-sought substance.[1]

Kunckel's formula involved the use of gold chloride and depended upon two factors that had eluded previous alchemists and craftsmen: first, the full tint is achieved only when an extremely small quantity of gold is present; and second, the color is not developed until the glass is reheated — on first cooling it is an almost-colorless gray. Once this process was understood, gold was eventually replaced by a suboxide of copper and, later, selenium, both of which produced even richer shades of red.[2]

Aside from its distinctive red color, this covered cup reveals another decorative element that was made possible by chemical and technological advances. The method used to engrave the woodland scene on the clear glass panel in the front of the cup was introduced into Germany by Italian rock-crystal carvers in the late sixteenth century. Its application to glass, however, did not become truly practical or popular until Bohemian glassmakers developed a good crystal glass that contained potash instead of soda and a larger proportion of lime than had previously been used. The properties of this new composition allowed it to be easily engraved or carved with a diamond, a copper wheel, or abrasives.[3]

The optical properties of glass are also used in the design of this cup. A concave, medallion-shaped piece of clear glass is set into the back of the cup, opposite the engraved bas-relief. This acts as a lens, reducing the much larger scene of a stag and does drinking from a stream in a wooded area, so that it may be viewed in its entirety through the small crystal opening. At the same time, the highlights and shadows are slightly intensified, increasing the sense of depth. Although the principle is different, the effect is similar to that produced by a stereoscope.

1. Fritz Kämpfer and Klaus G. Beyer, *Glass: A World History, The Story of 4000 Years of Fine Glass-Making*, trans. and rev. Edmund Launert (Greenwich, Conn.: New York Graphic Society, 1966), pp. 163, 306, 311.
2. The use of copper was known to medieval stained-glass workers; its rediscovery is credited, in part, to Kunckel. Edward Dillon, *Glass* (London: Methuen and Co., 1907), pp. 133, 289; Kämpfer and Beyer, *Glass: A World History*, p. 311.
3. E. M. Elville, *The Collector's Dictionary of Glass* (London: Country Life, 1961), p. 105; Dillon, *Glass*, pp. 278, 283-84.

29 Designed by Louis Comfort Tiffany, American, 1848-1933. *Three-Handled Vase*. Gold iridescent glass, ca. 1900, H. 7-3/4 inches (19.7 cm.). Gift of Mrs. Robert I. Gale, Jr., Mrs Caroline Macnaughton, and Fred R. White, Jr. CMA 62.424

Though glass is generally considered a stable material, it can be vulnerable to environmentally-caused internal changes and subject to decay. Over a period of time the supercooled liquid-alkali silicates from which the glass is made tend to crystallize or devitrify, which causes the glass to become more brittle as this change takes place. The actions of moisture, oxygen, and carbonic acid in the atmosphere also change the character of glass, causing it to decompose by dissolving the alkali and leaving silicic acid in the form of minute scales. These scales produce the dull, iridescent surface commonly associated with ancient glass that has long been buried in the ground.[1]

Toward the end of the nineteenth century the excavation of ancient glass objects in the Middle East generated a great deal of interest in this natural, chemical and physical phenomenon. The lustrous surfaces of these objects appealed to the fin-de-siècle taste and the predilection of that period for irregularity and richness of detail. As a result, many American and European artists sought means of recreating that silky texture and opalescent sheen. Perhaps none were more successful than Louis Comfort Tiffany, the foremost American representative of the Art Nouveau movement.

The satiny, mottled surface of the *Three-Handled Vase* [29] is typical of the deep translucency and lush metallic luster that characterize Tiffany glass. The vase also exemplifies the craftsman's understanding of his medium and the glassmaker's triumph over his materials. Tiffany's distinctive use of color and form are influenced by the nature of the glass itself. The graceful bell shape of the vase is a natural product of the blowpipe process. The subtle unevenness of the simple rim and the gentle curve of the thick, ropelike handles suggest the viscosity of the once-molten glass.

Depending on the angle of light, the shimmering colors blend into a rainbow of bluish-purple, rosy gold, or golden green tones, highlighted with turquoise. The trailing ivy motif that encircles the vase was created by working bits of colored glass into the still-molten body. This detail reinforces the sense of movement that enlivens the smooth surface. The subdued brown, green, and yellow tones of its matte finish effectively offset the sheen of the vase.

Tiffany thoroughly studied the chemistry and techniques of glassmaking, examining common preserve jars as well as ancient vessels and medieval cathedral windows. Though his experiments were successful, Tiffany left little specific information on the procedures used to create his distinctive glass.[2]

The August 1896 edition of the brochure *Tiffany Favrile Glass*, for example, declared in its addendum: "Mr. Tiffany obtains his iridescent and luster effects . . . by a careful study of the natural decay of glass, and by checking this process, by reversing the action in such a way as to arrive at the effects without disintegration."[3] The account went on to explain the chemical decomposition of glass, but omitted any explanation of the method used by Tiffany. Some insight may be gained, however, from the patent claim filed by Tiffany in 1880, a decade and a half earlier: "The metallic luster is produced by forming a film of a metal or its oxide, or a compound of a metal, on or in the glass, either by exposing it to vapors or gases or by direct application. It may also be produced by corroding the surface of the glass, such processes being well known to glass-manufacturers."[4]

Actually, Tiffany's method relied on gold chloride, both as a suspension within the glass and as a spray applied to its surface before cooling. A reducing flame brought the gold to the surface of the glass. The effect could be intensified by the use of a spray which etched the surface, creating a satinlike texture. This spray was made by dissolving twenty-dollar gold pieces in a solution of nitric and hydrochloric acid, which was then heated and thinned for use.[5]

29 Designed by Louis Comfort Tiffany, *Three-Handled Vase*.

1. Rutherford John Gettens and George L. Stout, *Painting Materials: A Short Encyclopaedia* (New York: D. Van Nostrand Co., 1942), p. 234.
2. Robert Koch, *Louis C. Tiffany: Rebel in Glass* (New York: Crown Publishers, 1964), p. 121; see also Joseph Purtell, *The Tiffany Touch* (New York: Random House, 1971), pp. 81-112; William B. O'Neal, "Three Art Nouveau Glass Makers," *Journal of Glass Studies* 2 (1960): 125-38.
3. Koch, *Rebel*, p. 50.
4. Ibid., pp. 121-22.
5. Ibid., p. 122.

30 Japan, Edo Period (1615-1867). *Wine Ewer with Cover.* Lacquer, mother-of-pearl, metal, late 18th century, H. 6-1/4 inches (15.9 cm.). Gift of Stouffer Restaurant-Inn Corp. CMA 71.1065

Lacquer is a gummy, resinous substance obtained by collecting, straining, and refining the sap of the lac tree. This tree is native to China, but was cultivated as early as the sixth century in Japan, where a highly organized industry emerged. Lacquer has also been used in Ceylon, Burma, and other southeast Asian countries. When applied to leather, paper, basketry, textiles, or wood, a coating of lacquer dries to a surface so dense and hard that it is virtually impervious to water, alcohol, acids, and the elements. The sap of the lac tree was originally used for strengthening and preserving purposes, and as a tightening medium for sealing boxes, chests, and other containers. Lacquer was also used on wooden household utensils, buildings, the equipment of warriors, and the stiff ceremonial headdresses worn by court officials. In addition to utilitarian functions, lacquer also came to be used as a decorative medium. The brownish glutinous resin can be worked into a deep black or vermilion color with a glossy shine that adds a lively sparkle to the surface treated. The basic process used to create a piece of Oriental lacquer, such as the *Wine Ewer with Cover* [30] or the sheath of the seventeenth-century dagger [43], is straightforward. It demands, however, immense concentration; patience; and a precise, empirical understanding of chemistry and physics.

First, a wooden core is fashioned in the shape of the finished object. After careful shaping and smoothing, it is primed and then covered with several layers of cloth impregnated with a mixture of liquid lacquer and clay. Once the final shape is defined, many layers of lacquer are applied, and each is carefully dried and polished before the next is brushed on. After the desired number of layers have been added, a final coat of highly refined, colored lacquer is put

30 *Wine Ewer with Cover.*

on. The design is then added in relief, as inlay, in successive layers of lacquer and combinations of gold dust, silver, shell, wood, semiprecious stones, and mineral pigments. The lid of this wine ewer is adorned with small chips of precisely cut mother-of-pearl set in a grid pattern; assorted shapes of the cut shell seem to float in the wavelike motif at its base.

The surface has been sealed with a final coat of clear lacquer which was thoroughly dried before it was polished with a series of natural abrasives. A lacquering process such as this may require as many as fifty lacquer-layers of various mixtures; if properly made, the article will last almost indefinitely.

Making lacquer ware involves subtle chemical reactions and an understanding of the physical properties of the materials being used. The illusion of depth in a lacquered surface is caused by the reflection of light. This effect is influenced by the number of layers of lacquer-embedded metal dust, the fineness of the metallic dusts or grains, the colors, the density of the inlaid materials, the combinations in which the layers are built up, and the amount and type of polishing.[1]

1. Information on the subject of lacquer was drawn from these sources: Martha Bayer, *Catalogue of Japanese Lacquers* (Baltimore: Walters Art Gallery, 1970), pp. 9-20; Barbara Adachi, *The Living Treasures of Japan* (New York: Kodansha International, 1973), pp. 36-40; see also Kurt Herberts, *Oriental Lacquer: Art and Technique* (London: Thames and Hudson, 1962), pp. 249-60.

31 Greece, 6th century BC. *Black-Figure Amphora*. Painted terra cotta, H. 12-5/8 inches (32.1 cm.). Gift from J. H. Wade. CMA 29.979

32 Greece, 4th century BC. *Red-Figure Bell-Krater*. Painted terra cotta, H. 12 inches (30.5 cm.). Gift from J. H. Wade. CMA 24.534

Pottery has no single origin. Clay, the raw material of pottery, occurs naturally in almost every region of the world, and as early as Upper Paleolithic times man made use of its moist plasticity to model figurines. The hardening of clay by fire was known equally early, since it occurs naturally when a patch of clay is chosen for a hearth, and by 9000 BC clay figurines were being intentionally fire-hardened in the Middle East.

Scholars have theorized that the making and firing of clay vessels followed shortly after the making of figurines, perhaps as an outgrowth of fire-hardened, clay-lined storage basins such as those found in the floors of the earliest-known neolithic dwellings. As pots were found to be not only useful but also pleasing to the eye, and as potters discovered that firing their ware to higher temperatures and combining various minerals with the simple clay body increased its durability and enriched its color and textural qualities, ceramics developed into one of the most versatile pyrotechnical arts.[1]

The stately red-figure and black-figure vases of classical Greece are considered to be among the most handsome ceramic objects ever created. They are also excellent examples of how the artist understood and made use of often-very-subtle material properties. Greek vase painting reached its artistic and technical heights in the pottery workshops of Athens, enjoying its greatest popularity from about 550 to 450 BC. Stylistically, the imagery on the red-figure vase is more realistic than its flat, black silhouette-like predecessor, and reflects a growing interest in man and his activities. The human form is more anatomically correct and assumes more natural poses; emotion or mood is also more subtly expressed through gestures and details. In general, the human figure became a major subject rather than simply an element in an overall decorative pattern.

The technique used to decorate the well-proportioned *Black-Figure Amphora* [31] belonged essentially to the engraver's art. The horses and men were painted with a slip as silhouettes against the lighter, orange-red background of the clay body. (Slip, a fluid mixture of clay and water with color oxides added, is used for decorating pottery.) Physical or narrative details, such as the horses' muscles or the men's facial features and clothing, were added with incised lines that cut through the slip and sometimes even scored the vessel wall.

The technique employed on the *Red-Figure Bell-Krater* [32] reversed the black-figure method. The seated woman and winged god were outlined with slip, and then the entire background was painted in; similarly, details were painted on with dilute slip instead of being incised.

In both styles the striking contrast between the warm red and black tones adds much to the vessels' visual strength. The colors are the result of the presence of iron oxides in the very fine-grained clay body, and both were produced during the three-stage firing

52

process. During the initial oxidizing phase of the firing, both the vessel and the slip turned red. In the reducing phase which followed, both turned black. Then, in the final reoxidizing phase, the porous fired clay of the vessel again turned red, but the slip-painted portions could not reoxidze to a red color because the slip had sintered, or burned, which sealed off its black iron oxide from contact with oxygen in the air.

Chemically, the red ferric oxide (Fe_2O_3) that is present in both the clay body and the unfired slip becomes black ferrous oxide (FeO) when the oxygen supply is reduced. When the oxygen supply is increased again, the unpainted parts of the vessel absorb oxygen, becoming once again red ferric oxide. The slip-covered portions, however, have developed a quartz layer which does not allow the reentry of oxygen, and therefore remains black. The matte red patches apparent between the two warriors on the reverse of the bell-krater evidently mark areas where the slip was too thin to protect against reoxidation.[2]

1. Lindsay Scott, "Pottery," in *A History of Technology*, eds. Charles Singer, E. J. Holmyard, and A. R. Hall, (Oxford: Clarendon Press, 1955), vol. 1, *From Early Times to Fall of Ancient Empires*, pp. 376-77; Cyril Stanley Smith, "Materials and the Development of Civilization and Science," *Science* 148 (14 May 1965): 908.
2. For major studies of ancient Athenian pottery techniques, see G. M. A. Richter, *The Craft of Athenian Pottery* (New Haven: Yale University Press, 1923) or J. V. Noble, *Techniques of Painted Attic Pottery* (New York: Watson-Guptill Publications, 1965), pp. 31-33.

31 *Black-Figure Amphora.*

32 *Red-Figure Bell-Krater.*

33 Italy, 3rd quarter, 17th century.
Albarello. Tin-glazed earthenware, H. 9 inches (22.8 cm.). The Norweb Collection. CMA 66.219

The seventeenth-century *Albarello* [33], a drug jar which was produced in Italy since the thirteenth century, is made of a tin-glazed earthenware commonly called majolica (sometimes spelled maiolica). A well-crafted, highly decorative ware, majolica — like most pottery — was usually intended for daily use in the home, shop, or pharmacy. Drug jars like this one, as well as many other types, sizes, and shapes, held the liquid, paste, or dry preparations that were used in early medicine. This particular albarello was

33 *Albarello.*

used to store dried materials; the characteristic concave sides made it easy to hold, and instead of a lid, it was closed with a piece of cloth stretched across the top.[1]

The making of majolica involved the basic pottery processes, but as with all pottery, its distinctive character is the result of a unique blend of chemicals and techniques. The shaped and bisque-fired earthenware was dipped in a liquid glaze made of lead, tin oxide, and a silicate of potash derived from wine lees mixed with sand. The tin made the mixture opaque, and when dry it formed an extremely porous, matte-white surface. Pigments made of common metallic oxides such as copper, iron, antimony, and, on this jar, cobalt and magnesium, were brushed onto the dry surface. These were immediately absorbed by the surface and non-erasable. This accounts somewhat for the loose, spontaneous character of majolica motifs, such as the imaginative floral patterns and the simple line drawing of a man on this albarello. During the firing, the oxide colors became incorporated into the white tin glaze beneath. Sometimes a final, clear overglaze was applied for a uniform brilliance.[2]

1. Bruce Cole, *Italian Maiolica from Midwestern Collections*, exh. cat. (Bloomington: Indiana University Art Museum, 1977), pp. 8-11. From the very first, production of this ware was influenced by foreign sources, and the term majolica itself may be derived from the ware imported from Majorca; see also William Bowyer Honey, *European Ceramic Art from the End of the Middle Ages to about 1815* (London: Faber and Faber, 1949), vol 2, *A Dictionary of Factories, Artists, Technical Terms, et cetera*, s.v. "Italian maiolica and faïence," pp. 325-26.

2. Cole, *Italian Maiolica*, pp. 10-11.

34 China, Sui Dynasty (581-618). *Vase with Loop Handles: Northern Brown Ware*. Glazed stoneware, H. 8-5/8 inches (21.9 cm.). Gift of Dr. and Mrs. Sherman E. Lee. CMA 54.784

35 China, Chekiang Province, Yü-yo District, Five Dynasties — Sung Dynasty, 10th century. *Ewer: Yüeh Ware*. Glazed buff stoneware, H. 8-1/2 inches (21.6 cm.). Gift of John L. Severance. CMA 28.177

36 China, Hopei Province, probably Ch'ü-yang District, Chin Dynasty, 12-13th century. *Deep Bowl: Ting Ware*. Glazed porcelain, D. 6-1/2 inches (16.4 cm.). Gift of Mr. and Mrs. Ralph King. CMA 17.383

37 China, Southern Sung-Yüan Dynasty (13th-14th century). *Seal Box: Ch'ing-pai Ware*. Glazed porcelain, D. 2-3/8 inches (6 cm.). Bequest of Elizabeth B. Blossom. CMA 72.256

Many complex changes are effected in clay by firing. These changes depend on the composition of the clay, the temperature of firing, the rate at which that temperature is achieved, and the gases that come in contact with the pot during all stages in the firing.[1] The deep bowl [36] and the seal box [37], for example, which are made of porcelain, have similar clay bodies and are closely related glaze types, but because of firing conditions their surface colors and textures differ greatly from each other. A similar contrast between the stoneware vase [34] and the ewer [35] is also evident.

The wide-mouthed, flaring-necked ewer [35], incised with leaf arabesques and floral rondels, was fired in an oxygen-reduced atmosphere. It was made from a rough, brownish buff stoneware body and coated with a high-iron-content glaze that fired into a shiny, dark, gray-green. The slightly yellow, heavily crazed patches (areas marked with a mesh of fine cracks) on the reverse of the ewer indicate where the firing heat was insufficiently controlled.

The vase with loop handles [34], made from a coarse, buff-colored stoneware, was also covered with a high-iron glaze. Unlike the ewer, however, it was fired in an oxygen-rich atmosphere, and therefore produced a different surface texture and color, with the matte, olive-brown glaze covering the vessel's flaring, cup-shaped mouth; narrow neck; and broad shoulders.

The interior motif of the thinly molded deep bowl [36] consists of a continuous band of lotus leaves and flowers, flowering water weeds with curling leaves, and a pair of ducks in flight. The interior base is decorated with two fish, one above the other and facing in opposite directions, on a ground of waves. The bowl's exterior is unadorned. The body is made of a whitish porcelain covered with a clear, ivory-toned glaze. A white slip coating between the clay body and the glaze further enhances the whiteness of its surface. The glaze had a low iron content and the bowl was fired in an oxygen-rich atmosphere, creating its warm tone.

The circular seal box [37], with a raised relief motif of floral sprays and leaves, was also coated with a low-iron glaze, but in

34 *Vase with Loop Handles: Northern Brown Ware*

35 *Ewer: Yüeh Ware*.

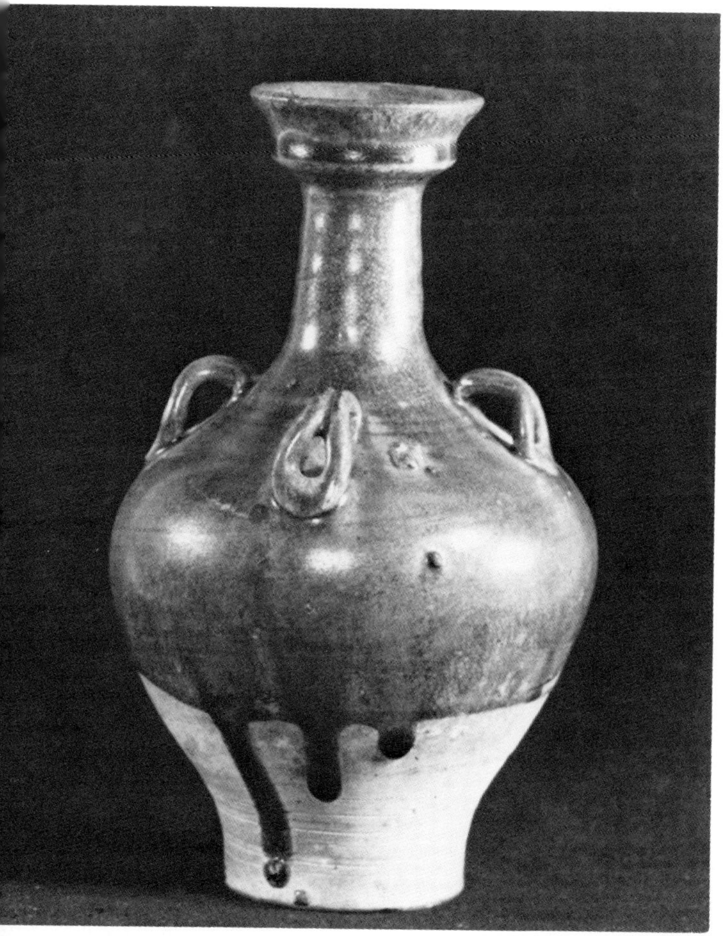

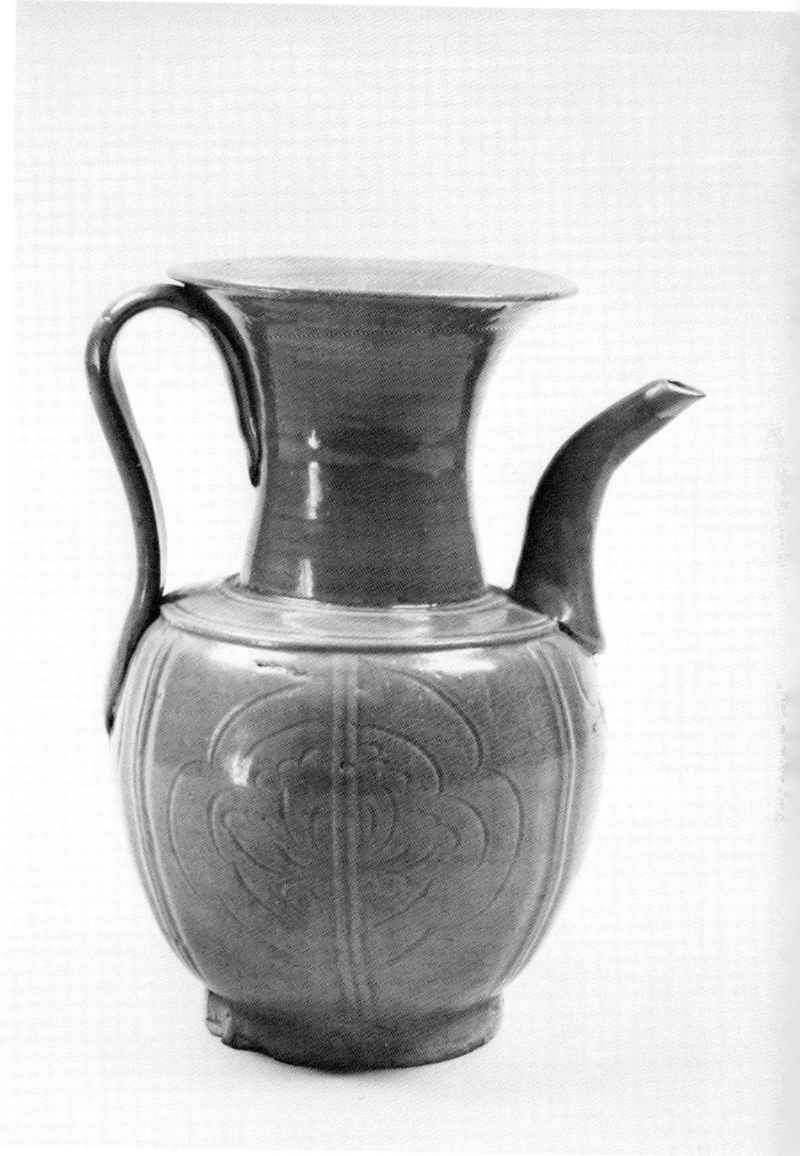

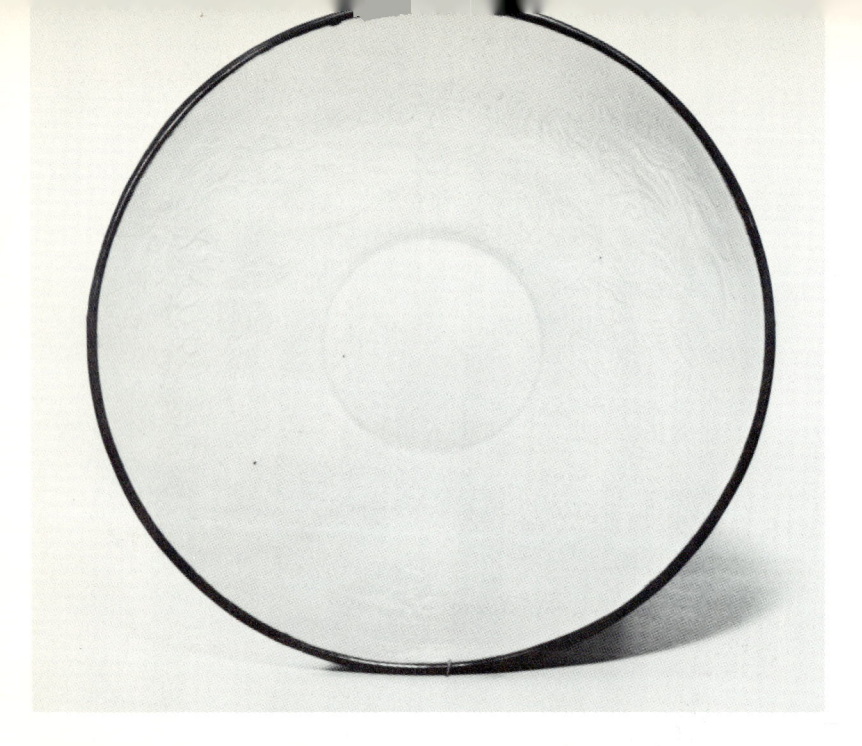

contrast to the deep bowl it was fired in a heavily oxygen-reduced atmosphere. This produced the somewhat opaque, greenish-blue surface color known as celadon. The soft reflection of light that creates the rich opalescent quality associated with celadon comes from the countless tiny bubbles trapped within the mature glaze. The bubbles, probably of water vapor, formed naturally as the object cooled in the kiln: surface tension causes their shapes to scatter light coming from any angle.

The production of glazed porcelain developed in China during the Shang dynasty, the same period during which the famous cast-bronze vessels emerged (see 40, text). The two technologies were closely related. Porcelain vitrifies, or matures, within the same range of temperatures necessary for the smelting of bronze. The use of clay piece-molds in the casting process also gave the Shang artisan an understanding of the behavior of clays at high temperatures; the potter readily applied this knowledge to the making of porcelain.[2]

1. Lindsay Scott, "Pottery," in *A History of Technology*, eds. Charles Singer, E. J. Holmyard, and A. R. Hall (Oxford: Clarendon Press, 1955), vol. 1, *From Early Times to Fall of Ancient Empires*, p. 382; Daniel Rhodes, *Clay and Glazes for the Potter*, rev. ed. (Philadelphia: Chilton Book Co., 1973), pp. 3-12.
2. Information concerning these pieces of pottery was drawn from Henry John Kleinhenz, "Pre-Ming Porcelains in the Chinese Ceramic Collection of The Cleveland Museum of Art" 2 vols. (Ph.D. dissertation, Case Western Reserve University, 1977), 1:42, 93, 135, 178; 2:604.

36 *Deep Bowl: Ting Ware.*

37 *Seal Box: Ch'ing-pai Ware.*

38 After Josiah Wedgwood, English, 1730-1795. *Copy of the Portland Vase.* Black and white jasper, after 1795, H. 10-3/8 inches (27.3 cm.). Bequest of Cornelia B. Warner. CMA 47.519

Josiah Wedgwood (1730-1795), the eighteenth-century English ceramist, was an artist, technologist, and scientist. The distinct ware that bears his name expresses the neo-classical attitudes of his time as well as his personal interest in antiquity. It also reflects the discoveries and industrial innovations of a practical, experimental chemist. Chemical science in Wedgwood's day was in a very early stage of development, and the penetrating observations and many systematic experiments recorded by him contributed to its growth.

The testing of materials and formulas, with a view to their use in clay bodies, glazes, and colors, constituted the greatest part of Wedgwood's experimental work. He also invented a pyrometer capable of measuring high degrees of temperature, and investigated the action of chemical reagents on clay bodies and at different temperatures. He applied his findings directly to the making of ceramic ware, and managed the production of his pottery as a fully integrated industry from research through sales distribution. He was friendly with such scientists as Erasmus Darwin and Joseph Priestley, and his notebooks continue to be a major source of information on the relationship between industry and science in the eighteenth century.[1]

38 *Copy of the Portland Vase.*

After years of trial, Wedgwood developed a unique high-fire stoneware body called "jasper," which could be stained with ordinary mineral oxides to tones of blue, lavender, green, or black. Barium sulfate was the principal and vital ingredient in the jasper body; when fired, the ware acquired a natural matte or nearly matte finish and a white opacity that could vary greatly. Sometimes it exhibited a dry, chalklike quality, but the finer varieties possessed a delicate hue and faint transparency.[2]

Wedgwood and his artists made use of this translucent character, forming from the white jasper delicate cameo reliefs. The color of the darker ground showed slightly through the thinly carved areas, creating a fine, light texture and enhancing the spatial effect. This translucency is quite apparent in the angel's wings and the upper tree leaves in the center of the primary relief on the *Copy of the Portland Vase* [38], made of black and white jasper ware.

The Portland Vase, the common name given to a remarkable ancient Roman glass amphora uncovered during the seventeenth century and finally housed in the British Museum, was first reproduced by Wedgwood in jasper ware in 1790.[3] A few years after Wedgwood's death a second series of the jasper-ware Portland Vase was produced, on which the cameo decoration is a molded application rather than carved as it was on the original. The vase in the exhibition is an example from a subsequent undated series.[4]

1. A. H. Church, *Josiah Wedgwood: Master-Potter* (New York: MacMillan and Co., 1903), pp. 63-64.
2. Originally, the entire clay body of jasper ware was colored by metallic oxides; later, only the surface was stained. The former ware is referred to as "solid jasper" the latter, "jasper-dip." Church, *Josiah Wedgwood*, pp. 36-28.
3. The original amphora is made of very dark blue glass, overlaid with a nearly opaque white glass from which cameo figures were cut in relief. Church, *Josiah Wedgwood*, pp. 30-31. See also William Burton, *Josiah Wedgwood and His Pottery* (New York: Cassell and Co., 1922), pp. 67-75.
4. Cyril Stanley Smith, *Aspects of Art and Science*, exh. cat. (Cambridge: Margaret Hutchinson Compton Gallery, Massachusetts Institute of Technology, 1978), p. 21.

39 Egypt, 27th-31st Dynasty (525-333 BC). *Bowl with High Neck*. Bronze, D. 5-1/2 inches (13.9 cm.). Gift of The John Huntington Art and Polytechnic Trust. CMA 14.581

Almost ten thousand years ago, not long after the development of agriculture made permanent communities possible, Middle Eastern man began using copper. At first the use was only occasional and confined to small objects of personal adornment, but it marked the beginning of metallurgy.[1]

A few metals, such as copper and gold, occur naturally as recognizable, malleable lumps on the earth's surface, but many more are found only in the form of mineral compounds, such as oxides, carbonates, or sulfides. Man could not make general use of metals, therefore, until he could systematically extract them from their ores. This process of extraction, known as smelting, developed about 3200 BC in Persia and Afghanistan, when it was discovered that the green stone, malachite, subjected to fire, produced the red metal, copper. Man's range of capabilities now increased dramatically: he possessed a material that could be hammered, drawn, cast, or molded into an ornament, a statue, a vessel, or a tool. It could even be thrown back into the fire and reshaped. But copper has one shortcoming: it is a very soft metal.[2]

Like every metal, copper is made up of layers of crystals; each layer is like a wafer in which the atoms of the metal are laid out in a specific pattern. In pure copper the crystals have parallel planes, so they easily slide past each other when the metal is stretched or bent. This weakness in the natural material was overcome, however, when the ancient smelters found, probably by accident, that by combining the even softer metal tin with pure copper they made an alloy that was harder and more durable than either: bronze.

Man has used bronze since its development about 2400 BC, but only in the last fifty years has science learned that the tin acts as a kind of atomic grit, adding points of roughness to the copper's crystal structure and preventing the layers from slipping. The presence of tin also checks the molten copper's absorption of oxygen and other gases, which makes bronze a better metal for casting than the native copper.[3] The Egyptian craftsman who fashioned the well-balanced bronze *Bowl with High Neck* was not aware of the complex crystalline structure that accounts for the strength and malleability of bronze, but he clearly understood how to manipulate these unseen factors.

The bowl, which was possibly used as a drinking vessel, began as a bronze ingot that was hammered into a flat sheet. The metal was further strengthened and tempered by repeated annealing, and then the half-hemisphere-shaped body of the vessel was formed by striking innumerable hammer blows in successively widening circles. This process, known as raising, caused the metal to flow rapidly outward as the circumference turned inward. By continuously rotating and skillfully hammering the pliable metal, the craftsman controlled the flow and deformation of the bronze. In this manner he also fashioned the bowl's sharply angled shoulder, gracefully flaring neck, and wide rim. The two narrow lines encircling the base of the neck and shoulder of the bowl, which emphasize its elegant, sweeping shape, were incised after the bowl was completed. Much of the surface is covered with a thin red skin of undetermined origin, but the remaining area is bright gold. This is a result of the high content (almost 12 percent) of tin in the bronze alloy. Tin in such a concentration produces a gold-like surface and also resists corrosion.[4]

1. Copper derives its name from Cyprus which, for the Romans, was the source of this metal. *Aes cyprium* means "bronze from Cyprus." H. J. Plenderleith and A. E. A. Werner, *The Conservation of Antiquities and Works of Art: Treatment, Repair, and Restoration*, 2nd rev. ed. (New York: Oxford University Press, 1971), p. 245.

2. Jacob Bronowski, *Ascent of Man* (Boston: Little, Brown and Co., 1973), pp. 125-26; R. F. Tylecote, *A History of Metallurgy* (London: The Metals Society, 1976), pp. 5-13.

3. Bronowski, *Ascent*, pp. 125-26; A. Lucas, *Ancient Egyptian Materials and Industries*, 4th ed., rev. and enl. by J. R. Harris (London: Edward Arnold, 1962), p. 217.

4. The Cleveland Museum of Art, Ancient Department, Curator's File; Cyril Stanley Smith, *Aspects of Art and Science*, exh. cat. (Cambridge: Margaret Hutchinson Compton Gallery, Massachusetts Institute of Technology, 1978), p. 12.

39 *Bowl with High Neck.*

40 China, late Shang Dynasty, 12th-11th century BC. *Fang-yu* (Square wine bucket). Bronze piece-mold casting, H. 10-1/2 inches (26.7 cm.). Purchase, John L. Severance Fund. CMA 63.103

The working of bronze reached its finest expression in ancient China. Artistic design and empirical knowledge have seldom been more closely related than in the making of the ceremonial bronzes of the late Shang dynasty. The material qualities of the metal itself were the main factors determining the shape and surface of the *Fang-yu* (square wine bucket) [40]. The vessel reflects at the same time the artist's mastery of bronze-working techniques.[1]

Separate piece-molds, which are slabs of clay, were impressed around a clay model approximating roughly in shape and surface design the bronze vessel to be cast. The piece-molds, bearing negative impressions of the design, were refined by carving and incising their leather-hard forms. After baking, these molds were reassembled around a plain clay core, probably the original model with its outer decorated surface scraped away. Molten bronze was poured from a crucible into the cavity between the core and piece-mold assembly. The walls of a vessel thus cast were about as thick as the clay scraped from the model and bore the molds' design in negative form, i.e., the model's original design in positive form. On this bucket, for example, stylized long-tailed birds are cast in high relief, their bold con-

40 *Fang-yu*.

tours rising above the fine-lined background of tight, squared spirals (*lei-wen*). These different levels of relief would have been exactly reversed in the molds, the birds deeply carved in intaglio and the spirals lightly incised.

The ribbed flanges that accent the corners of the container were an aesthetic solution to the technical problem of aligning and interlocking the piece-molds. Small pieces of bronze seem to have been used to separate the core and mold assembly, but molten metal inevitably leaked into cracks between adjacent mold segments. In early Shang bronze vessels, these leaks produced unsightly fins, but by late Shang, this problem was resolved by skillfully adjusting the mold-assembly to produce uniform flanges along the vertical joins. Flanges thus became an important feature of the design, not only accentuating the symmetrical composition, but also exposing the structural tension of the form.[2]

Although the amount of tin in bronze may range from 5 to 20 percent, the proportions of copper and tin used by the ancient Chinese were fairly exact; the best Shang bronzes contain approximately 15 percent tin. At that proportion the sharpness of the casting is perfect and the bronze is almost three times harder than copper. Lead was added to the bronze to increase the fluidity of the molten metal, to improve its casting qualities, and to soften the appearance of the final surface.[3]

Until very recently the *Fang-yu* was sealed by its lid, preserving the original warm, coppery-bronze color of the metal on the interior. However, long exposure to soil and water during burial have caused corrosion products to form a continuous film or patina over the exterior and in patches around the rim. These corrosion products, composed of oxidized minerals, have been identified in the Museum conservation laboratory: cuprite (red orange), tenorite (black), malachite (dark green), and azurite (bright blue).

1. Wai-Kam Ho, "Shang and Chou Bronzes," *The Bulletin of The Cleveland Museum of Art* 51 (September 1964): 174-87; for an extensive discussion of this tradition, see: Jon Eklund, "Art Opens Way for Science," *Chemical & Engineering News* (5 June 1978): 30; Jacob Bronowski, *The Ascent of Man* (Boston: Little, Brown and Co., 1973), pp. 126-28.
2. Wilma Fairbank, "Piece-mold Craftsmanship and Shang Bronze Design," *Archives of the Chinese Art Society of America* 16 (1962): 10-12; Noel Barnard, *Bronze Casting and Bronze Alloys in Ancient China* (Canberra: The Australian National University and Monumenta Serica, 1961), p. 117; Noel Barnard and Sato Tamosu, *Metallurgical Remains of Ancient China* (Tokyo: Nippon International Press, 1975), pp. 53-58.
3. Eklund, "Art Opens," p. 30; Bronowski, *Ascent*, p. 128; Cyril Stanley Smith, *Aspects of Art and Science*, exh. cat. (Cambridge: Margaret Hutchinson Compton Gallery, Massachusetts Institute of Technology, 1978), p. 12; Rutherford John Gettens, *The Freer Chinese Bronzes* (Washington: Smithsonian Institution, 1969), vol. 2, *Technical Studies*, pp. 33-56.

41 India, Sunga Period (185-72 BC). *Center-Bead and Two Triratna-Shaped Necklace Pendants.* Gold repoussé and granulation; Pendants, H. 2-1/4 inches (5.7 cm.); Bead, L. 2-1/4 inches (5.7 cm.). Purchase, John L. Severance Fund. CMA 73.66-73.68

The exquisitely detailed gold bead and pendants [41], made in India during the first or second century BC, are magnificent examples of two metallurgical skills — repoussé, a hammering technique, and granulation, a soldering process. The two side pendants are in the shape of the *triratna* (three jewels), a common Buddhist symbol of the Sunga Period.[1] The bead is barrel-shaped, with a central band of turtles. The shape of the small jewels and their intricate surface designs are inseparable from the techniques used to make them — techniques which are themselves inseparable from the innermost properties of the metals employed.

The easily deformed thin gold sheets were first raised into the general form of the pendants. The repoussé motifs, the turtles and the details of the *triratna*, were then raised from the back with hammer-struck, round-ended punches so that they stood out in relief. A uniform surface of very fine gold granules was soldered onto the jewels, entirely covering the background with a fine texture. Coarser granules, soldered on top of the initial layer, create simple flower shapes on the end pendants and an intricate web-like pattern on the central one.

The minute spherical gold granules were formed from small bits of pure gold which were heated and rolled about inside a vessel until they became solid and dropped through a hole into charcoal dust or water. They were then positioned on the jewels with an extremely thin coating of copper oxide. When heated, the reduction action of the furnace gases converted the copper oxide into metallic copper. This combined with a little of the underlying gold to form a low-melting-point alloy. Capillarity caused the molten alloy to flow into even the smallest crevices, forming nearly invisible bonds. Excess solder was removed with acidic mixtures.[2]

1. The Buddhist *triratna* is comparable to the sign of the cross in Christian tradition. The three jewels stand for the three segments of Buddhism: Buddha, Dharma (doctrine), and Sangha (order).
2. Information for this entry was obtained from: Cyril Stanley Smith, "Materials and the Development of Civilization and Science," *Science* 148 (14 May 1965): 912; Henry Hodges, *Artifacts: An Introduction to Early Materials and Technology* (London: John Baker, 1964), p. 78; The Cleveland Museum of Art, Oriental Department, Curators' File.

41 *Center-Bead and Two Triratna-Shaped Necklace Pendants.*

42 Japan, Edo Period (1615-1867). *Teakettle*. Iron sand-mold casting, H. 9-1/2 inches (24.1 cm.). Gift of Western Reserve Academy. CMA 55.558

Iron, "metal from heaven" as the ancient Sumerians named it, first became known to man as it appeared in meteorites. It occurred in these in its metallic state, in association with small amounts of copper, cobalt, nickel, and other elements, but remained a rarity for centuries because of its limited supply.

Man-made iron, smelted from terrestrial ores, was a much later development than copper or bronze, because processing it required high temperatures. A piece of a tool found embedded in an Egyptian pyramid and dating from before 2500 BC is the earliest known example of its use, but iron did not become a common material until around 1500 BC, about the time of the superb Shang bronzes in China (see 40).[1]

Historically, iron has been reduced from its ores at a temperature below its melting point. The resulting metallic sponge was hammered to remove excessive slag and to weld the iron particles into a compact, ductile mass. Because of its malleable quality, it was most often heated and hammered into form as wrought iron. In the Far East the casting of iron took precedence over this forging process. Cast iron, which contains about 3 percent carbon, is relatively easy to melt and cast in molds. It reached its finest artistic expression in the kettles made for the Japanese tea ceremony,[2] similar to the small teakettle [42] made during the Edo Period.

42 *Teakettle.*

The Japanese iron-making process began with extracting the iron ore dust and nuggets from mountain soil. Workers washed the soil downstream, collected material that sank to the bottom, and then repeated the procedure. The recovered ore was then melted in a furnace stacked with huge amounts of charcoal. Whipped into roaring flames for three days and nights, the charcoal burned and the kiln collapsed. When cool, the iron was carried out with chains and prepared either for forging or casting.[3]

The fluid sculptural quality and the rich textural surface of the sand-cast teakettle give testimony to appreciation for materials that forms the basis of the Japanese aesthetic. The deeply pitted, rippled texture and warm matte surface of the black iron is reminiscent of the metal in its native state. Silver inlay, in delicate leaf and grass motifs on the smooth wrought iron handle and the highly polished, red and copper lid, contrast sharply with the ruggedness of the kettle itself.

An inscription on the outer surface alludes to the solitary human figure:

The white-haired old man
   blows out his breath.
The solitary gibbon
   is wafted away.

1. Jacob Bronowski, *The Ascent of Man* (Boston: Little, Brown and Co., 1973), pp. 131-32; H. J. Plenderleith and A. E. Werner, *The Conservation of Antiquities and Works of Art: Treatment, Repair, and Restoration*, 2nd rev. ed. (New York: Oxford University Press, 1971), p. 281.
2. Cyril Stanley Smith, "Materials and the Development of Civilization and Science," *Science* 148 (14 May 1965): 913-14.
3. Victor and Takako Hauge, *Folk Traditions in Japanese Art*, exh. cat. (New York: International Exhibitions Foundation in Cooperation with The Japan Foundation, 1978/79), p. 30.

43 Nyudo Kotetsu, Japanese, 1602-1667. *Dagger and Scabbard*. Steel, *shibuichi*, and lacquer, blade length 11-7/16 inches (28.5 cm.), overall length 16-3/16 inches (41.1 cm.). Bascom Little Estate. CMA 74.56

The hard, steel edge, the strong, resilient back, and the unique curve make the Japanese sword a cutting weapon unequalled in the history of metallurgy. As a product of the Japanese forging and tempering methods that reached their peak of development in the thirteenth century (one thousand years after iron swords were introduced to Japan from China), the sword is far more than a weapon: it is an expression of a cultural and philosophical ideal. The process of making a steel sword involves delicate control of composition and heat treatment, as well as expertise in forging. It also involves achieving two different and seemingly incompatible qualities: the sword must be flexible, but it must also be hard. In Japan every step of the process is governed by time-honored ritual; it demands both physical stamina and empirical knowledge of the materials.[1]

The making of these fine swords begins with *jigane* (basic sword-making iron which is forged into thick wafers). After the *jigane* is forged and refined in a process that involves repeated heating and hammering, the swordsmith and his assistants begin the complex and exacting folding process. Sometimes folded lengthwise, sometimes across, the metal block is repeatedly heated and hammered thin, as many as fifteen or twenty times, to make the steel skin of the blade. This folding process purifies and strengthens the metal. The multitude of layers that result from the folding (for example, more than one million layers from twenty foldings) produces the fine grain visible in the area between the tempered edge and ridge of the blade. Under a microscope this fine granulation reveals itself as the crystalline transition between the martensite and pearlite. Grains of martensite shine against a duller background as the size and extent of the fine pearlite network fluctuates, depending on the area's carbon content and the rate at which it has cooled during the tempering process.[2]

The softer iron that will be the core or backbone of the sword is forged in the same manner, starting with small bits of several types of iron, which are heated, hammered, and eventually folded five to eight times. The resilient core and the durable blade are both made of iron, but in the blade this iron is alloyed with a tiny amount of carbon (usually less than 1 percent) to make steel, and the variations within that ratio determine the steel's underlying properties. Finally, the steel skin is skillfully folded over the iron core and the layers are joined. The block is then hammered into the length of the finished blade and receives its final shaping.[3]

The sword is tempered by a dramatic heating and cooling process. A wet paste, made up of charcoal, clay, whetstone powder, and water, is spread on the blade, thickly near the ridge, but more thinly on the blade's surface, in order to insure that the quenching of the heated blade will harden the cutting edge but leave the main part of the blade as flexible as it was before tempering. The clay acts as local thermal insulation; interestingly, any patterns drawn in it along the blade will appear in the design, color, and grain of the tempered edge. After inspection and preliminary sharpening and polishing, the blade is sent to a professional sword polisher. His painstaking work with various stones and powders enhances the beauty of the forged steel and brings the edge to a keenness once described as sharp enough to go "through three human bodies in a single stroke."[4]

The front of the steel blade on the Japanese dagger [43] made during the Edo Period (1615-1868) is decorated with two raised dragons facing each other, with the one nearer the hilt entwined around *vajra*-handled sword. Impressed images on the reverse side portray the Buddhist deity Fudo Myo-o and two attendant boys.[5] The slightly curved blade is set into a sharkskin-covered handle wrapped with reed laces and fitted with two phoenix-shaped mounts made of *shibuichi*, a copper and silver alloy. The gold and black lacquer scabbard, decorated with an overall phoenix and cloud design, was made by an unknown craftsman. It is also fitted with decorative *shibuichi* mounts in the form of a phoenix, a turtle, a scaly dragon, and what appear to be clouds

43 *Dagger and Scabbard.*

or waves. The smaller steel utility knife concealed within the scabbard has a dragon-decorated hilt and an inscription on the blade.

1. For an extensive analysis and clear explanation of the metallurgical art and structure of the Japanese sword, see: Cyril Stanley Smith, *A History of Metallography: The Development of Ideas on the Structure of Metals Before 1890* (Chicago: University of Chicago Press, 1960), pp. 40-62; idem, *A Metallographic Examination of Some Japanese Sword Blades, Estratto dal Symposium su La tecnica di fabbricazione delle iame di acciaio presso gli antichi* (Milan: Centro per la Storia della Metallurgia A. I. M., 1957). Barbara Adachi, *The Living Treasures of Japan* (New York: Kodansha International, 1973), p. 44.

2. Adachi, *Living Treasures*, pp. 43-45; Cyril Stanley Smith, *Aspects of Art and Science*, exh. cat. (Cambridge: Margaret Compton Hutchinson Gallery, Massachusetts Institute of Technology, 1978), p. 13; idem, *A History*, p. 41.

3. Adachi, *Living Treasures*, p. 45; Jacob Bronowski, *The Ascent of Man* (Boston: Little, Brown and Co., 1973), pp. 132-33.

4. Adachi, *Living Treasures*, p. 45.

5. The *vajra* is a symbol of power or authority. The sword blade as the central prong of this *vajra* emphasizes its symbolism. The deity Fudo Myo-o, an esoteric Buddhist god, is often pictured carrying a *vajra*-handled sword.

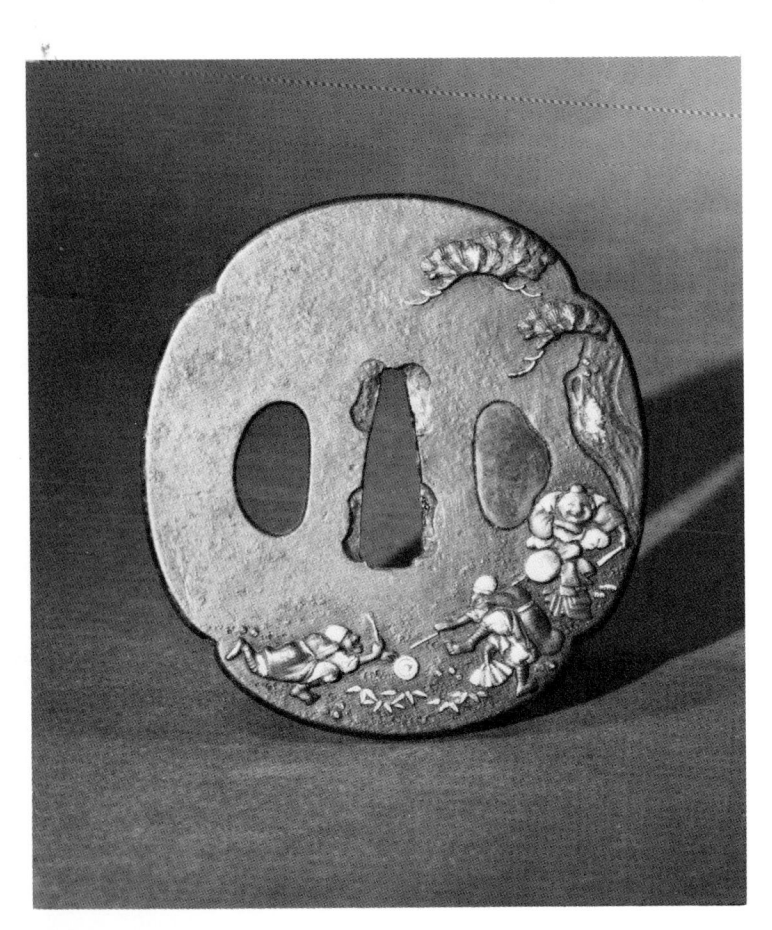

44 Japan, Edo Period (1615-1867). *Tsuba* (sword guard). Iron, gold, mother-of-pearl, H. 2-15/16 inches (7.5 cm.). Gift of D. Z. Norton. CMA 19.363

45 Japan, Edo Period (1617-1867). *Tsuba* (sword guard). Iron, H. 3-1/16 inches (7.4 cm.). Gift of D. Z. Norton. CMA 19.397

Tsuba, the protective shields which separate the handle and blade of a sword, are among the most exquisite examples of Japanese metallurgy. In their rich, varied textures and subtle blends of materials, they reveal the craftsman's appreciation of the structure of metals and their chemical properties.

Japanese metallurgists developed several alloys solely for the beautiful color or lush texture that they acquired after pickling, but natural imperfections — produced by scaling, corrosion, or forging accidents — were as highly prized as man-made effects. Roughly finished or superficially pitted pieces of iron were often used alone, or, as in the four-lobed *tsuba* [44], as a background for carefully fashioned gold and mother-of-pearl inlay. This guard is embellished with a relief design of three men sporting beneath a pine tree. On the reverse side are a tree in raised relief and an etched *torii* gate.[1]

Mokumé (wood grain), a most attractive surface effect that combines fine texture and coarse pattern, depends totally on the natural flow lines of a piece of metal during forging. The wood-grain-like patterns on the circular guard [45] were made by welding a rolled-up sheet of iron into a solid cylinder which was then flattened into a disk. After it was cut to shape, the metal was deeply etched to reveal its different layers; thus, the macrostructure of the iron became the design. In other *mokumé* guards, four or more iron or steel rolls were welded side by side and compressed end-wise; some were decorated with inlay while others were encrusted with small ornaments. Miochin Nobuiye (1486-1564), a metallurgist and an outstanding *tsuba* maker, is believed to have developed the *mokumé* technique.[2]

1. A Japanese *torii* is a port-and-lintel gate which usually stands before a Shinto shrine.
2. Cyril Stanley Smith, *A History of Metallography: The Development of Ideas on the Structure of Metals Before 1890* (Chicago: University of Chicago Press, 1960), pp. 57-58; idem, *Aspects of Art and Science*, exh. cat. (Cambridge: Margaret Compton Hutchinson Gallery, Massachusetts Institute of Technology, 1978), pp. 13, 42,

44 *Tsuba*.

45 *Tsuba*.

46 France (?), ca. 1850-1890. *Seated Figure, "Inspiration,"* possibly St. John the Evangelist. Copper, H. 12-1/2 inches (38 cm.). Collection of Professor and Mrs. George Levitine, Silver Spring, Maryland.

Electroplating and electroforming are related methods of depositing a coat of metal onto a base by means of electrolysis. Electroplating is a procedure in which a layer of metal, as thin as one-millionth of an inch or as thick as two-thousandths of an inch, is deposited on another metal. The electroforming process used in making the *Seated Figure, "Inspiration"* [46] is similar, except that the metal buildup is considerably thicker, and instead of a metal base the core is made of a disposable material such as wax, wood, metal foil, or plastic, and may be removed after the process of depositing the metal has been completed.

In both techniques the buildup is done by electrolysis — passing a low voltage direct current through an electric conducting core (cathode) suspended in a chemical solution (electrolyte) which contains a strip of metal (anode) that dissolves and forms a layer of metal on the core, or model of the finished object. The anode wears away gradually, at a rate that depends on the thickness of the metal deposited.[1]

Artists' use of electrolysis predates by many decades the discovery of voltaic electricity. Medieval craftsmen used an acid cupriferous solution to produce a copper coating on iron, and in the seventeenth century the industrial use of electrolytic replacement led to a minor art known as Herrengrund ware. This ware was made from the pure copper that was recovered from mine waste waters by means of electrochemistry. Pieces of scrap iron were placed in acidic mine waters containing dissolved copper. The copper ions in the water underwent an exchange process with the atoms of iron, which produced a soft deposit of copper.[2]

The duplication of coins and small art objects by electrodepositing began in 1838, and by the mid-nineteenth century it had become a popular practice that contributed to a widespread understanding of electricity. Gold and especially silver, which was deposited on nickel silver (a white copper alloy), soon became favored plate surfaces. Electroplating quickly became an important art industry that furnished economic support for the fledgling nickel industry and provided a thriving market for electromechanical generators before the invention of electric light. Developments in electrochemistry, along with telegraphy, also added to the growing body of scientific information that would give rise to electrical engineering.[3]

1. Sharr Choate with Bonnie Cecil De May, *Creative Gold- and Silversmithing* (New York: Crown Publishers, 1970), pp. 265, 273.
2. Cyril Stanley Smith, *Aspects of Art and Science*, exh. cat. (Cambridge: Margaret Compton Hutchinson Gallery, Massachusetts Institute of Technology, 1978), p. 13; idem, "Art, Science, and Technology: Notes on Their Historical Interaction," in *Perspectives in the History of Science and Technology*, ed. Duane H. D. Roller (Norman: University of Oklahoma Press, 1971), p. 145.
3. Smith, "Art, Science, and Technology," p. 146.

46 Seated Figure, "Inspiration"

Conclusion

Though artist and scientist no longer share a common workbench, they do share the world — a world in which an ever-expanding technology puts its stamp on our daily lives. The proliferating new tools, new materials, new means of production now often serve as catalyst to the artist. A whole new mode of painting emerged, for example, when artists such as Morris Louis and Kenneth Noland used acrylic paints developed after World War II — quick-drying and water soluble — which could be applied directly to untreated canvas and deeply saturate its fibers, thus eliminating the need for primers. Sometimes artists, adopting an industrial product for their own purposes, have developed a process which industry, in turn, takes up for *its* purposes. This happened with styrofoam, a soft, synthetic material developed by Dow Chemical which can be shaped or carved into the most complicated forms. In the mid-1960s several artists began to use this new material as a substitute for wax in the casting of metal sculpture; this process of "foam casting" was later adopted by industry.

Neon tubing, laser images, plexiglass, inflated polyethyline tubes, television cameras and monitors, computers — artists have used the whole panoply of technology to push beyond old boundaries, at times, it seems, almost to redefine art, and certainly to alter the relationship between the viewer and the work of art. Sometimes there has been a cooperative reaching out between artists and technologists to nourish and even to foster on an organizational basis the links between them. Gyorgy Kepes, painter, designer, photographer, and teacher who has been intensely involved in establishing networks of communication between art and science, once defined the Center for Advanced Visual Studies at the Massachusetts Institute of Technology as ". . . a clearing house of artistic tasks that have authentic roots in our present condition . . . a laboratory where the most advanced technological tools can be tested for their applicability in the newly emerging scale of artistic tasks . . . and . . . a testing ground for the needed collaboration among artists, scientists, and engineers."[1]

In 1967 J. Wilhelm (Billy) Klüver and Robert Rauschenberg organized Experiments in Art and Technology, Inc. (E. A. T.) whose purpose, in Rauschenberg's words, was to encourage "the inevitable, active involvement of industry, technology, and the arts," to ". . . guide the artist in achieving new art through new techology and work for the professional recongition of the engineer's technical contribution within the engineering community."[2] E. A. T. designed the revolutionary Pepsi-Cola Pavilion at the 1970 World's Fair in Osaka, Japan, a spherical structure above whose exterior floated a man-made water cloud, and whose interior contained the first light-sound system ever designed for such a spherical construction, as well as a spherical mirror — the largest ever made — which cast "three-dimensional" reflections of viewers onto the high domed ceiling. Klüver himself described the Pavilion as "a work of art with its own unity and integrity, an unexplored theater and concert space, a recording studio for multi-channel compositions and a field laboratory for scientific experiments."[3]

Art, in a world transformed by science, will probably continue to seek encounters with science itself. Kepes has written: "We need to map the world's new configuration with our senses, dispose our own activities and movements in conformity with its rhythms and discover its potentialities for a richer, more orderly and secure human life . . . . The new world has its own dimensions of light, color, space, forms, textures, rhythms of sound and movement — a wealth of qualities to be apprehended and experienced." If we are to understand this landscape, says Kepes, we must "build the images that will make it ours."[4]

1. Quoted in Douglas Davis, *Art and the Future: A History/Prophecy of the Collaboration Between Science, Technology and Art* (New York: Praeger Publishers, 1973), p. 117.
2. *E.A.T. News* 1 (1 June 1967): 1, quoted in Davis, *Art and the Future*, p. 137.
3. Quoted in Davis, *Art and the Future*, p. 137-38.
4. Gyorgy Kepes, ed., Introduction to *The New Landscape in Art and Science* (Chicago: Paul Theobald and Co., 1956), pp. 19, 20.

Selected Bibliography

Ackerman, James. "Science and Art in the Work of Leonardo." In *Leonardo's Legacy: An International Symposium*, pp. 205-25. Edited by C. D. O'Malley. Berkeley and Los Angeles: University of California Press, 1969.

Arber, Agnes. *Herbals: Their Origin and Evolution, A Chapter in the History of Botany 1470-1670*. Cambridge: At the University Press, 1938.

Ashdown, Charles Henry. *European Arms and Armour*. New York: Brussel & Brussel, 1967.

Barnard, Noel. *Bronze Casting and Bronze Alloys in Ancient China*. Canberra: The Australian National University and Monumenta Serica, 1961.

Barnard, Noel, and Tamosu, Sato. *Metallurgical Remains of Ancient China*. Tokyo: Nippon International Press, 1975.

Birrell, Verla. *The Textile Arts: A Handbook of Weaving, Braiding, Printing, and Other Textile Techniques*. New York: Schocken Books, 1976.

Blanc, Peter. *The Artist and the Atom*. From the Smithsonian Report for 1951, no. 4082, pp. 427-39. Washington, D. C.: United States Government Printing Office, 1952.

Blunt, Wilfrid, assisted by Stearn, William T. *The Art of Botanical Illustration*. London: Collins, 1950.

Bronowski, Jacob. *The Ascent of Man.* Boston: Little, Brown and Co., 1973.

———. "The Discovery of Form." In *Structure in Art and in Science*, pp. 55-65. Edited by Gyorgy Kepes. New York: George Braziller, 1965.

———. *Magic, Science and Civilization.* New York: Columbia University Press, 1978.

Choulant, Ludwig. *History and Bibliography of Anatomic Illustration in Its Relation to Anatomic Science and the Graphic Arts.* Translated by Frank Mortimer. Chicago: University of Chicago Press, 1920.

Cole, Bruce. *Italian Maiolica from Midwestern Collections.* Exh. cat. Bloomington: Indiana University Art Museum, 1977.

Davis, Douglas. *Art and the Future: A History/Prophecy of the Collaboration Between Science, Technology and Art.* New York: Praeger Publishers, 1973.

Dillon, Edward. *Glass.* London: Methuen and Co., 1907.

Dunthorne, Gordon. *Flower and Fruit Prints of the 18th and Early 19th Centuries.* Washington, D.C.: Gordon Dunthorne, 1938.

Eklund, Jon. "Art Opens Way for Science." *Chemical & Engineering News* (5 June 1978): 25-32.

Fairbanks, Wilma. "Piece-mold Craftsmanship and Shang Bronze Design." *Archives of the Chinese Art Society of America* 16 (1962): 9-15.

Fries, Waldman H. *The Double Elephant Folio: The Story of Audubon's Birds of America.* Chicago: American Library Association, 1973.

Gettens, Rutherford John. *The Freer Chinese Bronzes.* 2 vols. Washington, D. C.: Smithsonian Institution, 1969.

Gettens, Rutherford John, and Stout, George L. *Painting Materials: A Short Encyclopaedia.* New York: D. Van Nostrand Company, 1942.

Grigson, Geoffrey, and Buchanan, Handasyde. *Thornton's Temple of Flora.* London: Collins, 1951.

Hauge, Victor, and Hauge, Takako. *Folk Traditions in Japanese Art.* Exh. cat. New York: International Exhibitions Foundation in cooperation with the Japan Foundation, 1978/79.

Herberts, Kurt. *Oriental Lacquer: Art and Technique.* London: Thames and Hudson, 1962.

Herringham, Christina J. *The Book of the Art of Cennino Cennini.* London: George Allen and Unwin, 1922.

Herrman, Rolf-Dieter. "Art, Technology, and Nietzsche." *Journal of Aesthetics and Art Criticism* 32 (Fall 1973): 95-102.

A History of Technology. 7 vols. Oxford: Clarendon Press, 1955-1978. Vols. 1-5, edited by Charles Singer, E. J. Holmyard, and A. R. Hall, 1955-. Vols. 6-7, edited by Trevor I. Williams, 1978.

Ho, Wai-Kam. "Shang and Chou Bronzes." *The Bulletin of The Cleveland Museum of Art* 51 (September 1964): 174-87.

Hodges, Henry. *Artifacts: An Introduction to Early Materials and Technology.* London: John Baker, 1964.

Honey, William Bowyer. *European Ceramic Art from the End of the Middle Ages to about 1815.* 2 vols. London: Faber and Faber, 1949.

Kämpfer, Fritz, and Beyer, Klaus G. *Glass : A World History, the Story of 4000 Years of Fine Glass-Making.* Translated and revised by Edmund Launert. Greenwich, Conn.: New York Graphic Society, 1966.

Kandinsky, Wassily. "Rückblicke" [Reminiscences]. Berlin: Herwarth Walden, 1913. Translated in *Modern Artists on Art*, pp. 20-44. Edited by Robert L. Herbert. Englewood Cliffs, N.J.: Prentice-Hall, 1964.

Kleinhenz, Henry John. "Pre-Ming Porcelains in the Chinese Ceramic Collection of The Cleveland Museum of Art." 2 vols. Ph.D. dissertation, Case Western Reserve University, 1977.

Koch, Robert. *Louis C. Tiffany: Rebel in Glass.* New York: Crown Publishers, 1964.

Laporte, Paul M. "Cubism and Science." *Journal of Aesthetics and Art Criticism* 7 (March 1949): 243-56.

Lawalrée, André, and Buchheim, Günther. *P. J. Redouté: Facsimile Prints Made from Mostly Unpublished Original Paintings by Pierre-Joseph Redouté.* Switzerland: Gesellschaft Sweizerischer Rosenfreunde; Pittsburgh: Hunt Institute for Botanical Documentation, 1972.

Lucas, A. *Ancient Egyptian Materials and Industries.* 4th ed. Revised and enlarged by J. R. Harris. London: Edward Arnold, 1962.

Mason, S. F. *Main Currents of Scientific Thought: A History of the Sciences.* New York: Henry Schuman, 1953.

Mayer, Ralph. *The Artist's Handbook of Materials and Techniques.* New York: Viking Press, 1977.

Morley, Henry. *Palissy the Potter: The Life of Bernard Palissy, of Saintes, His Labours and Discoveries in Art and Science: With an Outline of His Philosophical Doctrines, and a Translation of Illustrative Selections from His Works.* 2 vols. London: Chapman and Hall, 1852.

Needham, Joseph, with research assistance of Ling, Wang. *Science and Civilization in China.* 5 vols. Cambridge: At the University Press, 1954-.

Noble, J. V. *Techniques of Painted Attic Pottery.* New York: Watson-Guptill Publications, 1965.

Olszweski, Edward J., with Glaubinger, Jane. *The Draftsman's Eye: Late Italian Renaissance Schools and Styles.* Cleveland: Cleveland Museum of Art, 1980.

The Original Water-Color Paintings by John James Audubon: For The Birds of America. Introduction by Marshall B. Davidson. 2 vols. New York: American Heritage Publishing Co., 1966.

Plenderleith, H. J. and Werner, A. E. A. *The Conservation of Antiquities and Works of Art: Treatment, Repair, and Restoration.* 2d ed., rev. New York: Oxford University Press, 1971.

Rhodes, Daniel. *Clay and Glazes for the Potter.* Rev. ed. Philadelphia: Chilton Book Co., 1973.

Rhys, Hedley Howell, ed. *Seventeenth- Century Science and the Arts.* Princeton: Princeton University Press, 1961.

Richards, L. S. "Drawings by Battista Franco." *The Bulletin of The Cleveland Museum of Art* 52 (October 1965): 107-12.

——————. "Giovanni Battista Franco's Anatomical Drawings in Cleveland." *Journal of the History of Medicine and Allied Sciences* 20 (1965): 406-9.

Richter, G. M. A. *The Craft of Athenian Pottery.* New Haven: Yale University Press, 1923.

Sarton, George. *A History of Science: Ancient Science Through the Golden Age of Greece.* Cambridge: Harvard University Press, 1952.

Sheon, Aaron. "French Art and Science in the Mid-Nineteenth Century: Some Points of Contact." *Art Quarterly* 34 (Winter 1971): 434-55.

Sitwell, Sacheverell, and Roger Madol. *Album de Redouté.* London: Collins, 1954.

Smith, Cyril Stanley. "Art, Technology and Science: Notes on Their Historical Interaction." In *Perspectives in the History of Science and Technology,* pp. 129-65. Edited by Duane H. D. Roller. Norman: University of Oklahoma Press, 1971.

——————. *Aspects of Art and Science.* Exh. cat. Cambridge: Margaret Hutchinson Compton Gallery, Massachusetts Institute of Technology, 1978.

——————. *A History of Metallography: The Development of Ideas on the Structure of Metals Before 1890.* Chicago: University of Chicago Press, 1960.

——————. "Materials and the Development of Civilization and Science." *Science* 148 (14 May 1965): 908-17.

——————. "Matter Versus Materials: A Historical View." *Science* 162 (November 1968): 637-44.

——————. *A Metallographic Examination of Some Japanese Sword Blades.* Estratto dal Symposium su la tecnica di fabbricazione delle iame di acciaio presso gli antichi. Milan: Centro per la Storia della Metallurgia A. I. M., 1957.

Smith, Cyril Stanley, and Hawthorne, John G. "Mappae Clavicula: A Little Key to the World of Medieval Techniques." *Transactions of the American Philosphical Society* 64 (July 1974).

Waddington, C. H. *Behind Appearance: A Study of the Relations Between Painting and the Natural Sciences in This Century.* Cambridge: M. I. T. Press, 1970.

Phone Renewal: 321-7621
DATE DUE

Access Services
DePaul University Library
2350 North Kenmore Ave.
Chicago, IL 60614

DE PAUL UNIVERSITY LIBRARY
30511000152860
701.05R476S C001
LPX SCIENCE WITHIN ART CLEVELAND, OHIO